David Christopher Davies

A Treatise on Earthy and Other Minerals and Mining

David Christopher Davies

A Treatise on Earthy and Other Minerals and Mining

ISBN/EAN: 9783744790468

Printed in Europe, USA, Canada, Australia, Japan

Cover: Foto ©berggeist007 / pixelio.de

More available books at **www.hansebooks.com**

WORKS BY THE SAME AUTHOR.

A TREATISE ON METALLIFEROUS MINERALS AND MINING. Illustrated with numerous Wood Engravings. Second Edition, revised. Crown 8vo, 12s. 6d. cloth. (Uniform with, and forming a companion volume to, " A Treatise on Earthy and other Minerals and Mining.")

"Without question, the most exhaustive and the most practically useful work we have seen; the amount of information given is enormous, and it is given concisely and intelligibly."—*Mining Journal.*

"The author has gathered together from all available sources a vast amount of really useful information. As a history of the present state of mining throughout the world this book has a real value, and it supplies an actual want, for no such information has hitherto been brought together within such limited space."—*Athenæum.*

"The volume is one which no student of Mineralogy should be without."—*Colliery Guardian.*

A TREATISE ON SLATE AND SLATE QUARRYING, SCIENTIFIC, PRACTICAL, AND COMMERCIAL. Illustrated with numerous Wood Engravings and Folding Plates. Second Edition, revised. 12mo, 3s. 6d. cloth.

"Mr. Davies has written a useful and practical handbook on an important industry, with all the conditions of which he appears familiar."—*Engineering.*

"One of the best and best-balanced treatises on a special subject, having at once scientific, practical, and commercial relations, that we have met with."—*Engineer.*

CROSBY LOCKWOOD AND CO.,
7, STATIONERS' HALL COURT, LONDON, E.C.

A TREATISE ON

ARTHY AND OTHER MINERALS AND MINING

BY

D. C. DAVIES, F.G.S.

MINING ENGINEER; AUTHOR OF "A TREATISE ON SLATE AND SLATE QUARRYING," "A TREATISE ON METALLIFEROUS MINERALS AND MINING," ETC., ETC.

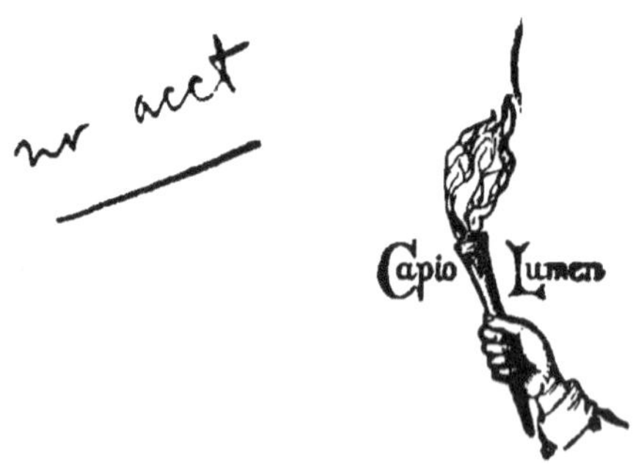

LONDON
CROSBY LOCKWOOD & CO.
7, STATIONERS' HALL COURT, LUDGATE HILL
1884

PREFACE.

THE design of this book is to give such a full and intelligible an account of the minerals selected for description in it as shall be useful and interesting to persons engaged in mining pursuits, and in the manufacture of the substances described; while it is also hoped that it will possess some interest for the general reader. It does not aspire to be a manual of mineralogy, although it is hoped that the classified list of minerals, with a brief description of each species, given at the end of the book, will serve to show the mineralogical position of each mineral described, and form an introduction, for the student who may desire it, to the more elaborate and systematic treatises on mineralogy. It is intended as a companion volume to my work on "Metalliferous Minerals and Mining," and perhaps, in point of time, it should have preceded that volume.

It will be observed that in the present book the distinctively earthy minerals are first considered. Next come those minerals which are compounds of earths and alkalies, some of which have a metallic base; after which carbon and compounds of carbon are noticed. Sulphur next occupies a place by itself, and then follow a number of metallic minerals distinguished by the extreme difficulty with which their metals have been extracted from them, and the rapidity with which those metals, when so obtained, and when unalloyed with others, unite with oxygen.

The student is thus led up to the series of useful and noble metals described in the volume on "Metalliferous Minerals and Mining." In both volumes prominence is given to the way in which the minerals described occur in the strata of the earth.

Of necessity a work of this kind must to a large extent be a compilation. It may, however, be permitted to me to say that a considerable portion of the information it contains is the result of my own observation and experience in the course of mining travels and work at home and abroad. I have thus been able to supplement the researches of others by my own.

My hope is that the work which is the outcome of these combined researches may prove a useful one, and, with its companion volume, be found of permanent value for information and reference.

D. C. DAVIES.

EBNAL LODGE, GOBOWEN, OSWESTRY,
April 15th, 1884.

CONTENTS.

PART I.—EARTHY MINERALS.

CHAPTER I.

SILICA AND SOME OF ITS COMBINATIONS.

CHAPTER II.

ALUMINA, MAGNESIA, LIME, WITH SOME OF THEIR COMBINATIONS.

CHAPTER III.

CLAYS.

PART II.—HALOID MINERALS.

CHAPTER IV.

SODIUM, CHLORINE, CHLORIDE OF SODIUM (COMMON SALT).

CHAPTER V.

CHLORIDE OF SODIUM—continued.

CHAPTER VI.

NITRATE OF SODA, BORAX, BARYTA, GYPSUM, FLUOR SPAR, AND ALUM SHALE.

CHAPTER VII.

PHOSPHATE OF LIME.

CHAPTER VIII.

PHOSPHATE OF LIME—continued.

CHAPTER IX.

PHOSPHATE OF LIME—continued.

CHAPTER X.

PHOSPHATE OF LIME—continued.

PART III.—CARBON, COMPOUNDS OF CARBON, AND SULPHUR.

CHAPTER XI.

CARBON AND CARBONACEOUS SUBSTANCES.

CHAPTER XII.

CARBON—continued.

CHAPTER XIII.

CARBON—continued.

CHAPTER XIV.

SULPHUR.

PART IV.—METALLIC MINERALS.

CHAPTER XV.

ARSENIC.

CHAPTER XVI.

COBALT.

CHAPTER XVII.

MOLYBDENUM—ANTIMONY.

CHAPTER XVIII.

MANGANESE.

CHAPTER XIX.

CLASSIFIED LIST OF MINERAL SUBSTANCES.

LIST OF ILLUSTRATIONS.

PART I.

EARTHY MINERALS.

SILICA, ALUMINA, LIME, MAGNESIA, GLUCINA, ZIRCONIA, THORIA, WITH SOME OF THEIR COMBINATIONS.

A TREATISE ON ARTHY AND OTHER MINERALS AND MINING.

CHAPTER I.

SILICA AND SOME OF ITS COMBINATIONS.

Silicon—Oxygen—Silica—Description—Proportions in Rock Masses—Redeposited Silica in Cavities, Cracks, and Beds—Daubrée's Experiments on the Decomposition and Crystallisation of Varieties of Quartz, Vitreous, Chalcedonic, Jaspery—Rock Crystal and its Varieties—Chalcedony—Agates — Flint — Chert—Jasper — Opal—Analyses of Rocks containing much Silica and Alumina.

SILICA.

Of the sixty-four simple elements of which, as far as we know, the earth is composed, the most abundant are silicon and oxygen. These two, combined in the proportion 51·96 of silicon with 48·04 of oxygen, form silica, of which mineral it is estimated that two-thirds of the earth's crust is formed.

The true nature of silica began to be investigated in the year 1807; but it was not understood until a few years afterwards, when Berzelius extracted from it the simple element silicon, which, on combining with oxygen in the proportion just given, forms the white powder known as silicic acid or silica.

The simple element silicon has been obtained, by Wohler and Deville, in transparent crystals as hard as the diamond, to which they bear a certain exterior resemblance; also in metallic crystals imitating graphite, and also in a black non-crystalline powder.

The colour of pure silica, as seen in crystallised quartz, is white; but along with the combination of silicon and oxygen in its composition there is usually a small admixture of other substances, chiefly metallic oxides, iron, manganese, &c., as hereafter described, and these give to it various other colours, including those which make some of its varieties valuable as precious stones.

In hardness, silica ranks as 7,[1] and it may be easily distinguished in this respect from the fact that it cannot be scratched by an ordinary penknife. In a massive form it ranges from opaque to translucent, but in separate crystals it is transparent. It crystallises into several shapes, the common form being a six-sided column, capped by a pyramid of an equal number of sides, as shown in fig. 1.

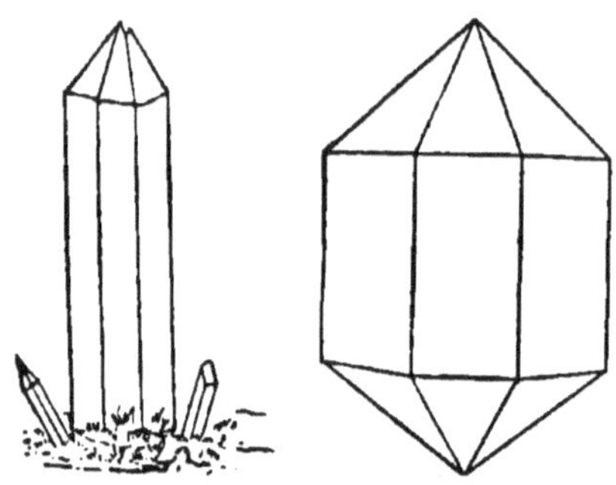

FIG. 1.—COMMON CRYSTALS OF QUARTZ.

Of itself it is infusible, but with soda it melts and forms glass. Under certain conditions, as will be seen, it is soluble. It forms the chief constituent of the rock masses of the globe, especially among the older rocks. In granitic and gneissic rocks it is present to the extent of 66 to 75 per cent. The imperfectly cleaved slaty rocks of the Cambrian and Silurian strata contain from 60 to 70 per cent. In the greenstone and syenitic rocks of the same formations it forms from 45 to 55 per cent. of the mass, and in the porphyritic rocks of the same groups it ranges from 59 to 75 per cent. In the sandstones of the millstone grit, the Coal-measures and of the New Red Sandstone group of rocks, it is present up to 92 per cent., the cementing matter consisting of small portions of lime, alumina, and magnesia. Full analyses of several of these rocks are given at the close of this chapter.

In these rock masses, besides alumina, which is in them the

[1] See Table on 'Hardness of Minerals,' in the concluding chapter of the book.

chief associate of silica, there are proportions of soda and potash up to 14 per cent., with small quantities of lime, magnesia, and the oxides of various metals. As we shall see in treating of clays, it is the presence of soda and potash which has, during vast periods of time, facilitated the dissolving of the silica out of the mass preliminary to its being redeposited with a portion of alumina as clays.

Besides this great redeposition of the silicious and aluminous portions of the older rocks in beds of clay, another of a finer and more delicate kind has also been going on, in which silica, variously coloured and combined, has been deposited in veins, cracks, and cavities of those rocks themselves, where, according to the age or other conditions of the rock, it forms, as quartz, the matrix of gold, copper, lead, zinc, or other metallic minerals; where, in other cracks, the silica has been closely packed, the result is an opaque quartz without much sign of crystallisation. Where, on the other hand, there has been space for the process, it has become crystallised into the beautiful transparent forms in which it is found.

Portions both of the massive quartz, and also of the crystals, are beautifully and variously coloured by the presence in different proportions of iron, copper, manganese, titanium, and other minerals, and these form the source of many of our precious stones, which, broken off their parent rock, have been rolled and polished as pebbles in the sea, or have been deposited with less friction, and hence are found in a more perfect state in conglomerates and breccias near the sources whence they were derived.

Much light was thrown upon the way in which crystals of silica and alumina are formed naturally by a series of interesting experiments made by M. Daubrée, in the year 1857, resulting in the artificial formation of crystals of these minerals.

M. Daubrée had observed in the mineral springs of Plombières, the waters of which contain silicate of potash and soda, and have a temperature of 70° Centigrade, the formation of certain well-known silicates and other minerals usually found in the veins of older rocks. The masonry near the springs was often seen to be impregnated with hyalite (a sort of transparent silica identical with that found in basaltic rocks), and sometimes apophylite (silicate of potash and lime) appeared in nice crystals.

The question then arose with M. Daubrée, If hydrated silicates form slowly in mineral springs at a not very high temperature, may not anhydrous silicates be more quickly produced by the action of water at a high temperature?

To answer this question he began a series of experiments which answered it in the affirmative. The experiments consisted chiefly in submitting the different substances in the presence of water to a heat of 400° Centigrade for a month together, in a closed glass tube, protected by an iron case.

As glass formed part of the apparatus, it naturally occurred to him to determine first of all what result this treatment would have on glass itself. He found that at the above temperature, by the simple action of water, glass undergoes a complete decomposition.

It first becomes opaque, earthy, and fragile, resembling kaolin, then gradually and regularly swells and transforms itself into a host of minute crystals, which were found on examination to be wollastonite (silica 52, lime 48), and at the same time the alkalies of the glass were dissolved. Soon the silica was deposited in the form of crystallised quartz. When alumina was present, the phenomena were modified. When obsidian was acted upon in like manner, minute crystals of felspar were formed, resembling in the mass granular trachyte. Clay and kaolin, which had been previously purified by washing, on being submitted to similar treatment, resulted in felspar mixed with crystals of quartz. The presence of oxide of iron in the decomposition of the glass gave pyroxene instead of wollastonite. This resembled the natural crystals found in the Tyrol and Piedmont. The crystals were beautifully crystallised, and possessed both their green colour and transparency.

As the result of these experiments, M. Daubrée naturally concluded that most, if not all, the silicates found in the early crystalline rocks were formed by the influence of water at a high temperature, this temperature being, of course, much lower than that of the point of fusion of such silicates.

Let us now notice those varieties and combinations of silica which from their shape, colour, lustre, and transparency have been valued as precious stones.

Quartz.—I have already described the way in which this variety of silica occurs in veins, lodes, and reefs in the older rocks—massive, partly in a crystalline form, and crystallised usually in six-sided prisms, capped by a pyramid as shown in fig. 1. It also occurs in radiated and in granular forms. The clear white varieties are pure silica, but it is tinged yellow, rose, or of a smoky colour, and is, indeed, of all shades of colour through the presence of metallic oxides. It has been divided into the following three varieties, in which are included the precious stones to be described:—

I. VITREOUS VARIETIES.
II. CHALCEDONIC VARIETIES.
III. JASPERY VARIETIES.

I. In the first, or VITREOUS VARIETIES, we have—

1. *Rock Crystal.*—In shape and colour as described above. The crystals are usually rooted in a mass of quartz. It is probably the mineral described by the ancients as *Krustallos*, ice, whence the word 'crystal.' The crystals are found of small size among the mountains of Wales and Scotland, and, indeed, wherever the older granitic, slaty, or felspathic rocks are found.

The clearest and finest specimens are brought from the island of Madagascar, where they are frequently found in blocks ranging from 50 to 100 lbs. in weight. Fine specimens also come from Switzerland, and from Auvergne, in France. The rock crystals of this country were formerly known as British or Cornish diamonds. It was formerly sold at from 5*s.* to 20*s.* per lb. for the purpose of splitting and grinding into spectacle glasses, and it was also used for stones for lockets, seals, and rings. In the middle of the last century it was largely used for buckles and buttons, and many persons were employed in cutting and manufacturing it.

In describing the apatite deposits of Norway, Chapter VII., I notice how closely quartz and titanium are associated and interlaced with each other and with apatite, and some specimens of crystallised quartz are found enclosing slender needles and grains of titanium. These examples are known in France as *flèches d'amour* (love's arrows), as well as crystals of chlorite. These specimens are much valued, and are worked up into many ornamental articles.

2. *Amethyst* is quartz coloured violet and purplish blue, of different degrees of intensity, from the presence of the oxides of iron or manganese. It passes often in the same specimen through rose-coloured to pale red, and even colourless. It derives its name from a supposition that it possessed a charm against drunkenness. The amethyst occurs in veins in the older rocks, and it often forms the inner part of agates derived from the same sources. The most valuable are the amethysts which come from Ceylon and India; then those which are found in Brazil. Inferior kinds come from Germany, Spain, and Siberia.

3. *Rose quartz*, pink, red, and inclining to violet blue in colour. Occurs in fractured masses, and is imperfectly transparent. The colour is most permanent in moisture. Occurs at Ben and Rabenstein, in Bavaria.

4. *Smoky quartz.*—Quartz crystals tinted with a smoky colour, becoming sometimes black and opaque. The Cairngorm stone from the mountains of Aberdeenshire seems to be related to the two last varieties and the next to be described.

5. *Yellow or Citron quartz* or *False Topaz*, which is often set and sold for topaz, but from which it may be distinguished by the absence of cleavage in it. Occurs in light yellow translucent crystals.

There are, besides these, other varieties, as *Milky quartz*, *Aventurine quartz*, in which the crystals or mass contain numerous spangles of golden-yellow mica. The name is said to have arisen from the incident of a French experimenter dropping at a venture some copper in molten glass, which produced a similar appearance. Also *Ferruginous quartz*, in which, from oxide of iron, the crystals are yellow, brownish yellow, and red crystals.

II.—In the CHALCEDONIC VARIETIES we have—

1. *Chalcedony.*—Ranges in colour from white through grey, green, and yellow, to brown. The bluish varieties are sometimes called sapphire. It is translucent or semi-transparent. It occurs in stalactite, reniform, or botryoidal masses which have been formed in cavities in greenstones and others of the older rocks. Into these cavities, as into miniature caverns, water holding silicious matter has penetrated and deposited its solid contents, consisting almost exclusively of silica tinged by the presence of other minerals. Some of these cavities are several feet in diameter, and besides the colouring of the encircling mass there is often, in the interior of the concretions in them, cavities or central nuclea which contain sometimes as many as twenty-four different substances, as silver, iron pyrites, rutile, magnetite, tremolite, mica, tourmaline, topaz, with water, naphtha, and atmospheric air. The mineral occurs in some of the mines of Cornwall, in Scotland, Tyrol, Bohemia, and Hungary. Some of the crystals are of large size, one in a museum in Paris measuring 3 ft. diameter and weighs 8 cwt. Chalcedony was obtained in ancient times from the vicinity of Chalcedon, in Asia Minor, whence its name. It is now also obtained from Scotland, the Faroe Islands, Iceland, India, and Arabia.

2. *Agates.*—In this variety the colours are arranged in concentric undulatory and zigzag lines; in the latter case the specimens are known as *mural* or *fortification agates;* also in wavy bands as folds of drapery, and in moss-like representations, as in the mocha stone, from the presence of manganese. All these forms are sometimes seen in the same example if it is large enough, together with small kidney and pea shaped concretions, from the presence of oxide of iron. There is frequently also a commingling of the fine cloud-like masses of chalcedony with the forms of amethyst, jasper, and agate in the same specimens. Figs. 2 and 3 will give an idea of the great variety of ways in which these forms of quartz arrange themselves.

3. *Flint* consists of silica, which in a very fine condition has been separated from the surrounding rock, and which,

attracted to some organic or inorganic nucleus, and sometimes only to itself, has grown in successive layers or bands, often of

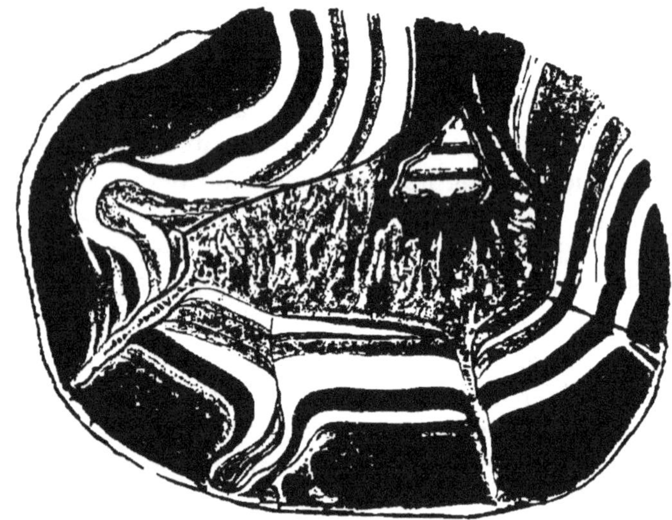

FIG. 2.—SECTION OF FOLDED AGATE.

different colours. An illustration of the way in which fine silicious particles become separated from the surrounding paste or mass is seen in the manufacture of pottery. When the

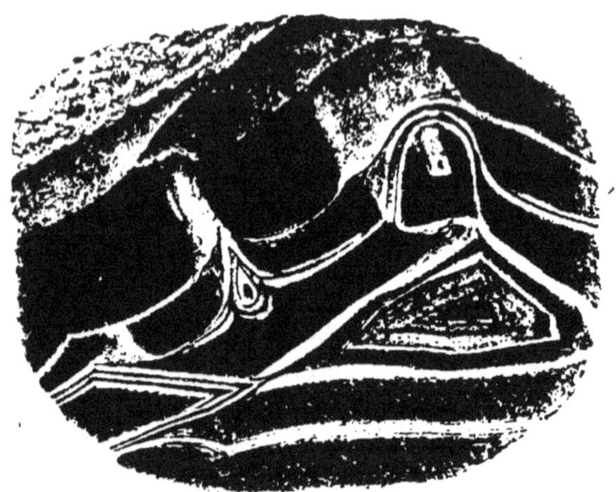

FIG. 3.—PERSPECTIVE VIEW OF FOLDED AGATE.

pounded or ground calcined flints are mixed with fine pipe-clay, and the mass is allowed to stand for a few days, the silica separates itself from the mass, and becomes aggregated into small nodules. In like manner have the layers of flints which occur so extensively in the chalk formation been separated from the limestone paste, and taken their present form. The occurrence of layers of flints at certain horizons of the chalk seems to indicate that silica was present in greater abundance during certain periods of the growth of the chalk beds than it was during others. Possibly also temperature and pressure may have had something to do with the matter. It has been seen that at a pressure of 60 lbs. in an alkaline solution and a high temperature, flint became perfectly soluble.

Very fine examples of layers of flints in the chalk beds may be seen in the range of chalk quarries extending at intervals from the city of Norwich to a point beyond the village of Thorpe. In some of these quarries I have seen the ancient trade of gun-flint manufacture carried on by men who looked as if they were survivors of the stone age themselves. Beautiful examples also of churches and houses built of these flints are to be seen in the city of Norwich, and indeed throughout the county of Norfolk generally.

4. *Hornstone*, or *Chert*, is allied to flint, but it is more brittle, and it takes its colour—dirty-grey, red, and reddish yellow, green, or brown—from the rocks in which it is found. It occurs in portions of sandstone rocks usually containing a little lime, the fine silica being seemingly collected into one spot. Calcareous portions of the beds of the millstone grit, with portions of rocks of similar composition in the oolite and greensand formations, show this structure.

Other chalcedonic varieties are *Onyx*, which consists of alternate horizontal layers, white, brown, or black in colour. *Sard* from the shores of the Red Sea, of a deep brown or blood red colour. Sard and white chalcedony combined form *Sardonyx*, a stone that was much used in ancient time for cameos, of which some beautiful specimens remain. *Chrysoprase*, apple-green in colour, from the presence of nickel. *Carnelian*, a clear, rich-tinted, bright red chalcedony, and *Cat's-eye*, composed of 95 per cent. of silica with minute proportions of alumina, lime, and oxide of iron, greenish grey, translucent, with a shining, vitreous, or resinous lustre, and when cut spherically giving the glaring internal reflections like the eye of a cat, which come from the presence of asbestos. It is obtained from Ceylon and the coast of Malabar.

It would be an interesting study to inquire into the nature of the operations which have resulted in the varied combinations and arrangements of colour and shape presented in the chalcedonic and jaspery variety of quartz, but it is beyond the scope of this work. I would refer the readers who feel an interest on the subject to a series of clear and beautiful papers, with no less beautiful illustrations, by John Ruskin, Esq., F.G.S., on brecciated and banded structures, which are contained in the fourth and fifth volumes of the *Geological Magazine.*

I may, however, say that from the foregoing and other considerations, the colours and the grouping of them, and the materials, whether partaking more of the nature of simple transparent quartz or fine cloudy chalcedony, with the shapes and positions occupied by each, would be affected by a number of circumstances, as the presence and proportion of metallic oxides, and the variations in this proportion at different periods during the long growth of the accretions in the mass, or the secretions in the cavity or fissure, by temperature, now right for the formation of crystals; and then, as in M. Daubrée's experiments, partially dissolving and rounding them, like the rounded apatite, pyroxene and other crystals in the Laurentian rocks, then surrounding the perfect or partially dissolved crystals with gelatinous matter. Then each mineral present would have its natural tendency to crystallise in its own way—oxide of iron in reniform shape, manganese in dendritic or moss-like forms, titanium in long thin prisms, quartz and alumina in their prevalent forms, and all these perhaps pressing upon and modifying each other. Then there would be times of drying and shrinkage, followed by an inflow or addition of pasty matter. While, during the partial dissolution of the crystals, the matter, becoming soft, would settle down in horizontal or other layers according to the foundation on which they rested or the nature of the force by which they were pressed.

III. JASPERY VARIETIES.

1. *Jasper*, a silicious rock of a hardened clayey nature, of a dull red or yellow colour. It has indeed been described as ferruginous clay. From the above colours it ranges through a great diversity, and often two or more colours are combined in the same specimen in bands, dots, stripes, and flames. Like chalcedony it occurs in nests, cavities, and concentric nodules. Striped green and brown jaspers from Siberia are much used, but the most valuable is the Egyptian jasper, in which the bands or ribbons occur in excentric zones, which are usually cut across to be polished. *Ruin* jasper presents the appearance of a group of ruins. *Porcelain* jasper resembles baked clay; it differs from ordinary jasper in that it is fusible before the blow-pipe. *Red porphyry* is like jasper in some respects, but differs from that variety in that it is fusible before the blow-pipe. Other varieties are *Bloodstone* or *Heliotrope*, *Lydian stone*, *Touchstone*, and *Basanite*.

2. *Opal* is a hardened paste or gelatine consisting of from 87 to 95 per cent. of soluble silica with from 5 to 13 per cent. of water. Its hardness is less than quartz, 5·5 to 6·5, and its specific gravity lighter, 2·21. Its usual colour is milk-white or pearl grey, and when looked through towards the light it presents with a milky transparence rose red and yellowish white, with a rich variation of colours as its position is changed—emerald and other shades of green, fire red, bright blue, violet, purple, and pearl grey. Sometimes the colours are arranged in small spangles; it is then called *Harlequin opal;* and when in broad plates or in wavy or flame-like delineations, the two favourite colours being rich orange yellow, when it is known as golden opal, and vivid emerald green. The colours are the more valued because they are produced by the remarkable power the mineral possesses of refracting the sun's rays. Opal occurs in veins in porphyritic rocks and in rolled fragments in drifted matter. The largest example known is in the imperial cabinet at Vienna; this weighs 17 ounces, and belongs to the variety known as *Precious* or *Noble opal.* Other varieties of opal are *Fire opal*, *Girasol*, *Common opal*, *Hydrophane*, *Hyalite*,

occurring in small glassy concretions. Opal is found in the Faroe Islands, near Freyburg, Saxony, Kaschan, Hungary, and in Honduras. The mineral was much valued by the ancients, who called it the "beautiful child of love." And the story is told of a Roman senator who gave up his life rather than resign an opal ring of great beauty to the Emperor Nero.

On the next page I give a table of analyses of various rocks into the composition of which silica and alumina largely enter, but chiefly the former. The higher the percentage of silica, the greater, it has been proved, is the power of the stone to resist the action of the weather. Since the table has been in type I have been favoured with the following analysis of the New Red Sandstone worked in the extensive quarries at Grinshill, near Shrewsbury :—

Silica	95·46
Alumina	1·17
Iron peroxide	0·87
Lime, carbonate	0·61
Magnesia, carbonate	·69
Water, combined	·91
Water at 212° F.	·77
	100·48

The mean crushing strain of this stone is 5,165 lbs. to the square inch. The amount of silica contained in several other sandstones used in building is as follows :—

Craiglieth.	93·3	Plean	95·64
Darley Dale	96·40	Rawdon Hill	92·825
Corsehill	95·24	Spinkwell and Clifford	88·5

Before noticing the rock masses of which silica forms the chief constituent, let us describe in the next chapter the mineral, alumina, with which it is so generally associated, especially in the older rocks of the earth's crust, and then, after referring to the metallic base, aluminium, and some of the forms in which combinations of alumina appear as precious stones, I will proceed to notice some of the features and characteristics of the rock masses composed chiefly of the two minerals, and record some particulars relative to the quarrying of the same.

ANALYSES OF VARIOUS ROCKS, INTO THE COMPOSITION OF WHICH THE MINERALS SILICA AND ALUMINA LARGELY ENTER.

—	Granite from Fox Rock, near Dublin.	Rose-coloured Porphyry from Greenville, Canada.	Syenitic Greenstone from ridge north of Portmadoc, North Wales.	Basalt from Fingal's Cave, Staffa.	Light Green Serpentine, from Galway.	Roofing Slate, Carnarvonshire, Wales (Cambrian Group).	Bluish Grey Building Stone, from the base of the Bala or Caradoc beds, Montgomeryshire.	Garth Trevor Stone, North Wales Millstone Grit.	Coal-measure Sandstone, Heddon, near Newcastle-on-Tyne.	Calcareous Red Sandstone, Mansfield, (Permian).	New Red Sandstone, Liverpool.
Silica	73.00	72·20	60·69	47·80	40·12	60·50	61·00	86·10	95·1	49·4	85·550
Alumina	13·64	12·50	18·50	14·40	2·00	19·70	31·00	4·20	—	—	7·570
Lime	1·84	0·90	5·41	10·16	—	1·120	1·45	·40	0·8	26·5	0·588
Oxides of Iron	2·44	3·70	4·73	12·30	3·47	7·83	1·60	8·00	2·3	3·2	2·668
Oxide of Manganese	—	—	trace	2·80	—	—	0·35	—	—	—	—
Magnesia	0·11	—	2·54	9·53	40·04	2·20	—	—	—	16·1	—
Soda	3·53	5·30 }	5·42	1·16	{ —	2·20	3·20	—	—	—	1·113
Potash	4·21	3·88 }			{ —	3·18	1.15	—	—	—	0·915
Carbonic Acid	—	—	—	—	—	—	—	—	—	—	—
Water and loss	1·20	0·60	2·71	3·00	13·36	3·30	0·25	0·07	1·8	4·8	1·596
	99·97	99·08	100·00	100·05	98·99	100·03	100·00	98·77	100·0	100·0	100·000

CHAPTER II.

ALUMINA, MAGNESIA, LIME, WITH SOME OF THEIR COMBINATIONS.

Aluminium—Alumina—Bauxite—Valued Forms of Alumina—Corundum—Sapphire—Ruby—Topaz—Brazilian Deposits of Emerald—Analyses of Beryl—Emerald Mines of Grenada—Tourmaline—The Precious Stones Deposits of Ceylon—More Massive Forms of Silica and Alumina—Orthoclase—Adularia—Felspar—Mica—Magnesia—Magnesium—Talc—Steatite—Chlorite—Serpentine—Pyroxene—Asbestos—Rock Masses—Granites and Gneiss Rocks—Syenitic and Dioritic Greenstones—Slaty Building Stones—Liverpool Corporation Quarry, Llanwddyn—Felspathic Rocks of North Wales—Lime and Limestones—Varieties of Costs of working Glucina, Zerconia, Thoria—Chemical Composition of various Limestones.

ALUMINA.

ALUMINA consists of aluminium and oxygen in the proportion of two parts of the former to three of the latter. In ordinary use alumina is a white powder, shapeless, infusible, and scarcely soluble. In a crystalline form it is found in the most perfect state as corundum. In an impurer state and mixed with other substances it becomes opaque, as in the case of emery, a common form. It is a mineral very abundant in nature, forming, as will be seen in the following analysis, from one-fourth to one-third of the substance of many of the older rocks of the earth's crust. Its specific gravity is 3·9. It is largely used in the processes of dyeing and calico-printing. *Aluminium*, the metallic base of alumina, is a light, whitish-coloured metal of bright lustre, which, as far as it has been worked, has been found very useful in the manufacture of optical and mathematical instruments, and for the lighter kinds of ornamental work. Its specific gravity is only 2·6. The extraction of the metal from its earthy surroundings has been carried on during the greater part of the last twenty years at Newcastle-upon-Tyne, with but indifferent success, the process being costly and intricate. Recently, a Birmingham manufacturer claims to have discovered a more simple and much cheaper process. Four parts of aluminium mixed with ninety parts of copper affords an alloy possessed of the greatest strength combined with malleability and ductility, and other alloys are being constantly adopted.

The material chiefly used for its production hitherto has been bauxite, a ferruginous clay obtained from Baux, near Arles, in the south of France. Its general composition is as follows :—

Alumina	57·4
Silica	2·8
Sesquioxide of iron	25·5
Oxide of titanium	3·1
Carbonate of lime	0·4
Water	10·8
	100·0

A similar material, containing from 44 to 54 of alumina and from 1 to 15 per cent. of iron, has also been used at Newcastle from the mines of the Irish Hill Company, Ireland.

It will be readily inferred that aluminium is one of the most abundant metallic minerals in nature.

To simplify its extraction from the clays and rocks in which it is contained is one of the greatest metallurgical problems of the present time. As this is solved many of the clays described in another chapter will become more valuable, and the metal more largely used.

Let us now notice some of the forms and combinations of alumina which on account of their beauty have been greatly valued.

Corundum is, as already observed, pure alumina in a crystalline condition. The forms are somewhat varied, but it occurs chiefly in six-sided prisms, as shown in fig. 4. It also occurs in a granular form. Its usual colours are blue and greyish blue, but it is also found red, yellow, and brown of various shades. Translucent to transparent. The specific gravity is from 3·9 to 4·16, and in hardness it ranks next to the diamond, scratching quartz easily. Except with borax it is infusible before the blow-pipe. Comminuted corundum occurs abundantly near Canton, and is much used in that city in the polishing of precious stones. Among the varieties of corundum are—

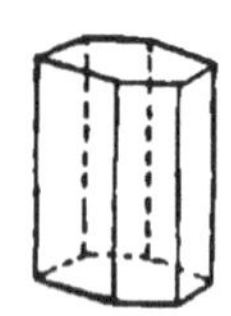

FIG. 4.—COMMON CRYSTAL OF CORUNDUM.

1. *Sapphire.*—The general composition of which is alumina 92, silica 5·25, oxide of iron 1·0. The colour most valued is a highly transparent bright Prussian blue. More frequently the colour is a pale blue, passing by paler shades into perfectly colourless varieties. The paler varieties are frequently marked by dark blue spots and streaks, which detract from their value. But these paler varieties lose their colour when subjected to great heat, a fact that has sometimes been taken advantage of by unscrupulous dealers to pass them off as diamonds.

The principal form of the sapphire is an acute rhomboid, but it has many modifications and varieties. On being broken it shows a conchoidal fracture, seldom a lamellar appearance. The best sapphires were formerly found chiefly in Ava and Pegu ; the paler varieties in the sands of rivers in Ceylon, inferior kinds being obtained from near Forez, in France. More recently the sapphire has been found in many localities in the United States of America. It belongs to the older gneissic and talcose rocks and granular limestones, and with fragments of these it is found in driftal deposits.

The sapphire is a gem prized next to the diamond. The largest known weighs about two ounces. There was also a fine rhomboidal crystal among the crown jewels of France, which weighed over an ounce.

2. *Ruby.*—The ruby is subdivided into several varieties according to colour, which in its turn is affected by mineral composition, *spinel* ruby occurring in bright red or scarlet crystals, *rubicelle* of an orange red colour, *balas* ruby rose red, *adamandine* ruby violet, *chlorospinel* green, and *pleonaste* is

the name given to dark varieties. The three next analyses show the difference in the composition of three spinel rubies.

1.		2.		3.	
Alumina . . .	90	Alumina . .	82·47	Alumina . . .	69·0
Silica	7	Magnesia. .	8·78	Magnesia. . .	26·2
Oxide of Iron .	1·2	Chromic acid	6·18	Silica	2
Goss	1·8			Oxide of iron .	0·7
				Chromic acid .	1·1
	100·0		97·43		99·0

The first of these is most nearly allied to corundum. Its hardness is slightly less than that of the sapphire, and it is infusible before the blow-pipe, except with borax, and then it is fused with difficulty. Its specific gravity is about 3·9.

The crystals are usually small, and when not defaced by friction they have a brilliant lustre, as has also the lamellar structure, with natural joints, which it shows on being broken. It exhibits various degrees of transparency. The colour most valued is the intense blood red or carmine colour of the spinel ruby. When the colour is a lilac blue, the specimen was formerly known as the Oriental amethyst, and was regarded as a connecting link between the ruby and the sapphire. Rubies are found in Pegu, in the sand of rivers near the town of Siriam. It is also found with the sapphire in the river deposits of Ceylon, and in various localities in the United States of America, in some of which the crystals have partly decomposed, and show a soft structure resembling steatite. In America it occurs in gneissic and metamorphic rocks, and in granular limestones.

3. *Topaz* derives its name from *topazo*, to seek, the mineral first known by this name being obtained from an island in the Red Sea which was usually surrounded by fog.

Two examples of the mineral now known by this name, gave on analysis the following results :—

1.		2.	
Alumina . . .	50·0	Alumina . . .	57·5
Silica . . .	29·0	Silica . . .	34·2
Fluoric acid . .	19·0	Fluorine . .	15·0
	98·0		106·7

It varies in size from two carats to three or four ounces. The specific gravity is 3·53.

By itself it is infusible before the blow-pipe. It possesses a brilliant transverse cleavage. It has a lustre greater than that of rock crystal. Fine topazes of a greenish yellow colour to perfectly white come from Siberia, Kamtchatka, and Australia. Pale greenish ones are found in the Highlands of Scotland, and small colourless examples come from St. Michael's Mount, Cornwall.

Topazes are found in large numbers in the neighbourhood of Villa Rica, in Brazil. They occur in small veins partially filled with talcose matter, and associated with quartz and specular iron ore. They are also found by thousands in the débris derived from the wearing down of granitic and gneissic rocks, perfect specimens being, however, rare. The searching for them and the preparation of them for sale is a considerable industry, and gives employment to a large number of persons.

The Brazilian topazes are of three kinds: blue, also called Brazilian sapphire; the yellow, of various shades of yellow; the deeper the colour, so that the stone retains its transparency, the more valuable it is. This on exposure becomes pink and red in colour, when it is known as the Brazilian ruby. These, with the white topaz, are found in a rolled, and more rarely a crystallised form, in the conglomerate described in the chapter on the diamond.

The topaz has sometimes been mistaken for the diamond. Apart from the suspicion that some supposed diamonds in royal collections are topazes, a notable instance of a mistake of this kind occurred in the year 1856. A topaz supposed to be a diamond was brought from Brazil, weighing about 189 carats, or about twenty-five ozs., and caused great excitement in Europe. It was estimated to be worth several million francs. At last a consultation of authorities was held in Vienna, the result of which was the statement—'The pretended diamond turns out to be a topaz, having the specific gravity and the hardness of an ordinary topaz, and is worth as a curiosity about

250 francs.' Common topaz is found abundantly in the vicinity of Falun, Sweden.

4. *Emerald. Beryl.*—The emerald is of a beautiful rich green colour, passing also into blue and yellow. It crystallises in six-sided prisms. It has a vitreous or resinous lustre, and varies from translucent to transparent. The following are the result of some analyses of the emerald and its varieties.

	Emerald from Mexico.	Emerald from Finland.	Emerald from Peru.
Silica	67·9	67·359	64·5
Alumina	17·9	16·465	16·6
Glucina	12·4	12·747	13·0
Magnesia	0·9	—	—
Oxide of iron	—	1·490	—
Soda	0·7	—	—
Titanic acid	trace	0·280	—
Oxide of chrome	—	—	5·25
Lime	—	—	1·06
	99·8	98·341	99·42

	Beryl, or Aquamarine.	Chrysoberyl, or Cymophane.
Silica	68	18
Alumina	15	71
Glucina	14	—
Oxide of iron	1	1·5
Lime	2	6
	100	96·5

The specific gravity is about 2·7, its hardness greater than that of quartz. Before the blow-pipe it is fusible into a grey and rather frothy glass.

The emerald was formerly obtained from Ethiopia, and was prized in ancient times. Necklaces of emeralds have been found in the ruins of Herculaneum. The chief source during the last three hundred years has been Peru, in the vice-royalty of Santa Fé; and in the valley of Tunca, between the mountains of New Grenada and Pompaya, they are found in veins traversing clay slate, and in cavities in certain granites. They are accompanied by quartz, calcareous spar, felspar, mica, and pyrites. The largest emeralds known are from Peru. They are about six inches long by two inches thick, but the largest specimens are seldom the purest.

A productive emerald mine is, or was, that of Muso, in New Grenada, Mexico. The emeralds occur in veins and cavities in a black limestone that contains fossil ammonites. The limestone also contains within itself minute emeralds, and an appreciable quantity of glucina. When first obtained the emeralds from this mine are soft and fragile; the largest and finest emeralds could be reduced to powder by squeezing and rubbing them with the hand. After exposure to the air for a little time they become hard and fit for the jewellers' use. In *Beryl*, or *Aquamarine*, the colour is a pale sea-green, passing on one side to light sky blue and greenish blue, and on the other into greenish yellow. Sometimes the same crystal presents two or more colours, and sometimes it is iridescent.

5. *Tourmaline* is composed of from 40 to 43 per cent. of silica, and about the same quantity of alumina, about 10 per cent. of soda, with up to 8 or 9 per cent. of manganese. It crystallises in prisms, with three, six, nine, or twelve sides. It is green in colour, ranging to blue, red, yellow, and brown. It is harder than quartz, and its specific gravity is from 3·0 to 3·3. It fuses before the blow-pipe into a spongy greyish white enamel. It is found in Siberia, Ava, and Ceylon, also in Brazil, where the stone is much worn in rings by the ecclesiastical dignitaries.

In the foregoing descriptions of precious stones, reference has been frequently made to Ceylon as one important source whence many of them have been derived.

The gems occur in an ancient gravel deposit, known as Nellan, which is frequently from ten to twenty feet below the surface. It consists of fragments and pebbles of granite, gneiss, and other of the older rocks imbedded in clay. It is covered by a hard crust a few inches in thickness, called Kadua, and which in places protects the underlying Nellan from the action of the streams. This is overlaid by recent gravel. In the Nellan there are large lumps of granite and gneiss in the hollows, as well as in pockets in the clay, which are known by the natives as elephants' footsteps. In and about these the precious stones occur in groups, where they are found by the miner.

The gem collector digs down to this stratum; he takes the clayey gravel out and places it by the side of his digging until he has accumulated three or four cubic yards. He then carries it in shallow basin-like baskets from two to three feet in diameter to a neighbouring stream, where he washes it until all the clay has disappeared, leaving only the sand containing gems behind. The gems are then carefully picked out, the washer removing with the palm of his hand one thin layer after another until the whole of the sand has been effectually searched.

The industry is not encouraged by the Government; inasmuch as it attracts a numerous loose population from agricultural pursuits and the more steady industries of the country. It is after all a poor trade, although now and then a lucky find is made. The chief town of the gem district is Ralnapoora, where most of the stones are polished.

The materials of the Nellan or gem drift, seem to have been derived from the wearing down of the large grained silicious granitic rocks that abound in parts of Ceylon as well as in Pegu, and along the coasts of China and Japan northwards. Results of this decomposition of these granitic rocks may be seen along the coasts of Japan in the loose sand that covers and gradually passes into the solid parts of the rock.

Other varieties of the combinations of silica and alumina with other substances as precious stones will be found in the concluding chapter, and we may now proceed to notice those combinations of the two minerals with others that contribute largely to the formation of the rock masses which, valuable in themselves as building and other stones, form also the depositories of the minerals, metallic and otherwise, described in this book and its companion volume.

Felspar.—Orthoclase is composed of silica 64·20, alumina 18·40, and potash 16·95. It crystallises in oblique rhombic prisms. Its common colours are white, grey, and pale red; but it also passes into greenish and bluish white. It has a vitreous and occasionally pearly lustre. Its hardness is 6, and it may be scratched with a good penknife. Specific gravity

2·39 to 2·62. It is not affected by acids, and it fuzes with borax into a transparent glass.

One of the finest varieties of felspar is that known as *Adularia*, from Mount Adula, near the St. Gothard Pass, where it is found redeposited from the rock mass in veins and cavities. It consists of silica 64, alumina 20, lime 2, and potash 14. *Moonstone* is another variety with bluish white spots of a pearly lustre. *Sunstone* is another, with a pale yellow colour, with minute scales of mica. *Aventurine* felspar, sprinkled with iridescent spots from the presence of minute particles of titanium or iron.

Mica.—Chemical composition: silica 46·2, alumina 36·8, potash 9·2, peroxide of iron 4·5, fluoric acid 0·7, water 1·8. In colour ranging from white to green, yellow, brown, and black. Pearly lustre, tough and elastic; occurs in thin plates or scales, and sometimes in radiated groups of the same. H. 2· to 2·5; gr. 2·8 to 3·. Mica differs from talc in not having the greasy feel of the latter, and in its thinner and more elastic plates; some of these occur of considerable size. They have been used in Siberia for glass; hence the name *Muscovy Glass*. Plates two or three feet diameter, and quite transparent, are found in New Hampshire.

We must now add to our list another of the earthy minerals, which, to a considerable extent, enters into the composition of the rock masses of the earth.

MAGNESIA.

Magnesia is a compound of magnesium and oxygen, in the proportion of 158 parts of the former to 100 parts of the latter. It is the only oxide of magnesium.

Magnesium.—This simple element is the metallic base of magnesia. It has in the metallic state the colour and lustre of silver; it is malleable, and fuses at a red heat, a little above which point it burns with great brilliancy, oxidizes, and forms magnesia. It also oxidizes on exposure to a moist atmosphere, but it is not affected in dry air.

Magnesia is a soft white powder, which is highly infusible.

It combines with water, but not so readily as lime. The artificial preparations of magnesia by precipitation from its soluble salts have the silkiness, lustre, and softness which are observed in asbestos, soapstone, and other magnesian minerals.

Among the minerals helping to form rock masses into which magnesia enters are the following :—

Talc.—Composition : silica 62·8, magnesia 32·4, protoxide of iron 1·6, alumina 1·0, water 2·2. In some examples the water amounts to 4 per cent. H. = 1 ; gr. = 25 to 29. Occurs usually in foliated masses made up of thin easily separable plates. It also passes into a crystalline, granular, and a fine impalpable texture. It has a pearly lustre, and, with most other minerals into which magnesia enters, it has an unctuous feel; colours—silvery white, greenish white, grey, green, and olive green. Some forty years ago it was much used in the manufacture of lamps and lanterns, more so than at the present time. It includes *Foliated Talc*, Soapstone, or *Steatite*, a massive variety of talc of a grey or greenish colour, and internally a crystalline texture; feels to the touch like soap. The composition of steatite is silica 62·2, magnesia 30·5, protoxide of iron 2·5. It is flexible, but not elastic like mica. Potstone, an impure talc, is another variety.

Steatite occurs abundantly in America. In small quantities it may be found in many rocks. From the facility with which it can be cut, drilled, and worked generally, and the polish it will take, it has been used for various internal portions of architecture. It is also used in the manufacture of porcelain, as a lubricant for machinery, and in the final polishing of the harder stones.

Chlorite.—Chemical composition : magnesia 34·0, silica 30·4, alumina 17, protoxide of iron 4·4, water 12·6. H. = 1·5, gr. 2·85. Occurs in masses of a dark olive green colour, and crystallises into hexagonal prisms. Occurs in thin plates and radiated forms like talc, pearly lustre, opaque to partly translucent. It is distinguishable from serpentine by a granular texture, and from talc by its yielding water in a glass tube, and from green iron earth by its extreme infusibility. Chlorite enters largely into the composition of schistose and slaty rocks.

Serpentine occurs in dark oil or olive green masses, also in a fibrous and lamellar form. These consist of thin plates or folia of a greenish white to dark green colour. H. 2·25 to 4·0; gr. 2·5 to 2·6. An analysis of a rock variety is given in the table, page 15, and the following is the composition of a purer variety from Cornwall, known as *Precious* serpentine: magnesia 44·2, silica 42·3, protoxide of iron 0·4, carbonic acid 0·9, water 12·4; capable of a high polish, and forms a beautiful stone. Becomes brownish red before the blow-pipe and gives off water. Another variety, named *Marmolite*, is brittle, but consists in easily separable thin folds. Its composition is: magnesia 41·4, silica 40·1, protoxide of iron 2·7, water 15·7. Serpentine of various kinds is worked in Cornwall and in America as ornamental stones, but it does not bear exposure to the weather.

Pyroxene occurs in various shades of green, passing towards white on one side, and brown and black on the other, yellow excluded. It has a vitreous lustre, inclining to resinous or pearly. In the massive varieties there is a coarse granular and sometimes fibrous structure, the fibres long and thin; crystallises in oblique rhombic prisms. The composition, as to the minor constituents, is somewhat varied, but the crystalline forms remain unchanged. H. = 5·6; brittle; gr. = 3·2, 3·5.

Pyroxene has been divided into three groups or divisions: the white or light coloured, the dark coloured, and the thin foliated.

I. *White Augite* or *Malacolite* includes several lesser varieties. Its general composition is: silica 55·3, lime 27·0, magnesia 17·0, protoxide of iron 2·2, protoxide of manganese 1·6.

II. *Augite* also includes several dark green varieties, in which there is a larger proportion of iron and manganese than in the first. The composition of one variety is given as— silica 54·1, lime 23·5, magnesia 11·5, protoxide of iron 10·0, protoxide of manganese 0·6.

III. The third class includes *Diallage*, *Bronzite*, and *Hypersthene*, all of which are characterized by being thin foliated. The composition of hypersthene is given as: silica 54·25, lime 1·5, magnesia 14·0, protoxide of iron 24·5, protoxide of manganese a trace, alumina 2·25, water 1·0.

HORNBLENDE.—Occurs in oblique rhombic prisms, long slender prisms, in columnar forms, and in fibrous masses of coarse and fine fibres, silken and like flax. In colour it ranges from white through bluish green, greyish green, green, and brownish green shades to black; a vitreous lustre, the faces of the plates or cleavage lines inclined from pearly opaque to transparent. H. = 5 to 6; gr., 2·9 to 3·4. Hornblende is divided into first, the light-coloured varieties, and second, into the dark-coloured varieties. The former include *Actinolite* and *Asbestos*, with the sub-varieties belonging to each. These varieties are distinguished by not containing much alumina or iron. The composition of glassy actinolite is as follows: silica 59·75, magnesia 21·1, lime 14·25, protoxide of iron 3·9, protoxide of manganese 0·3, hydrofluoric acid 0·8.

The dark varieties include *Hornblende*, the composition of which is: silica 48·8, magnesia 13·6, lime 10·2, alumina 7·5, protoxide of iron 18·75, protoxide of manganese 1·15, hydrofluoric acid and water 0·9.

Another variety is *Pargasite*, from Pargas, in Finland. This occurs in short thick crystals, and is composed as follows: silica 46·3, magnesia 19·0, lime 14·0, protoxide of iron 3·5, protoxide of manganese 0·4, hydrofluoric acid and water 2·2.

Asbestos.—Before leaving hornblende it may be well, on account of its rising commercial importance, to say a few words concerning this mineral. It was known to the ancients, who made of it the wicks for the lamps in their temples. These wicks served to feed the flame with oil up their fine fibres, but remained unconsumed; hence the name asbestos—unconsumed. The natives of Greenland now use it for the same purpose. Because it was easily cleaned the ancients gave it the name Amiantus—undefiled. It is now woven into cloth for packing the joints of steam-engines and machinery. It helps to form an excellent non-conducting cover for boilers. It is woven into fireproof garments, and for many other purposes. It occurs in large masses in felspathic and chloritic rocks. I have seen it in a coarse form associated with the greenish apatite-bearing rocks of South Norway. It is largely obtained at present from Italy for English manufacturers. It occurs in various forms—in slender flax-like fibres, and with a rich satin lustre, in seams in the rocks; in a hard and compact form of yellow and brownish colours, *Ligniform Asbestos;* in thin, tough sheets, like leather, *Mountain Leather*, which consists of thin beds or layers of matted fibres of asbestos; and in thicker masses of the same, *Mountain Cork.*

We may now proceed to notice some of the rock masses into the composition of which the minerals already described enter, as shown in the analyses given on p. 15.

Granites.—The particular composition of granites is given in the table. It is generally described as consisting of quartz, felspar, and mica, the first being pure silica, and the other two constituted as already described. There are, however, many variations, according as one or other of the minerals predominate, or the precise form in which it is present. Thus, felspar may be present as orthoclase, oligoclase, or albite, or two of these may be present. There may also be two kinds of mica present, and occasionally mica is replaced by hornblende. The crystals of felspar may be large and distinct, and the rock thus assume a porphyritic structure. These variations affect the colour. An abundance of flesh or pink-coloured felspar gives a reddish tint, like that of the granite of Peterhead. White felspar, or a preponderance of quartz, gives, with mica, a grey speckled stone, while hornblende imparts a greenish cast.

When the grains or particles are arranged in layers, the stone is called *Foliated Granite.* When this bedding becomes very distinct and the particles are fine, granite becomes gneiss, like those of Donegal and Galway, and the masses resting upon the older coarser granites of Norway and Sweden. Occasionally the particles of felspar are arranged in the quartz, or the quartz in the felspar, like the letters in Oriental writing, and then it is known as *Graphic Granite.* The names *Quartzose*

Granite, *Felspathic Granite*, and *Micaceous Granite* are given as one or other of the minerals predominate in the composition.

Granite was long considered to be of igneous origin, but the arrangement of the particles in layers, the presence of water in the quartz, with other considerations, have tended to modify this opinion. It would seem to be a sedimentary rock altered, and we may have granite in various stages of its history—in its original sedimentary form; when altered by heat; when protruded or injected from altered masses deep down in the earth through other rocks; or with the arrangement of its particles altered by pressure, similar to that which has produced the phenomena of slaty cleavage.

Generally speaking, granitic and gneissic rocks lie near the base, as far as this is known, of the geologic series. We see this in the position occupied by these rocks on the western coast of Great Britain, from St. David's up to Scotland, in the position of similar rocks in Ireland, in the position of the main bosses of granite in Devon and Cornwall, in the arrangement of the rocks all over the peninsula of Norway and Sweden, and in the place occupied by them in most of the great mountain chains of the world.

There are exceptions. The granites of the Alps and of the eastern Pyrenees are believed to be newer than the chalk. May it not be that these more recent granites, like some probably in Cornwall, are projected or intrusive granites, portions of the old vast underlying expanse of the ancient granites thrown up through and over the newer strata?

Granitic and gneissic rocks are not usually difficult to work. The component parts are hard, but the grain is open, and as a rule not more, if so much, is paid for boring or drilling and for driving tunnels or sinking shafts in them than is paid for the same work in the slaty rocks of Wales.

In Norway and Sweden the price paid for boring 1¼ holes is ¾*d.* per inch, or 9*d.* per foot. For driving a tunnel 7 ft. by 6 ft., 7*l.* 10*s.* per fathom; for sinking shaft 12 ft. by 8 ft., 9*l.* to 12*l.* per fathom; for open cuttings about 12 ft. wide, 25*s.* per cubic fathom. These were all in gneissic rock. The rate of wages being lower in these countries than in England, similar work would cost more here. Some of the close-grained granites in Scotland are difficult to drill, and it is found advisable to bring large masses of rock down at once by sinking a shaft or driving a tunnel, and putting a large quantity of explosive in them.

Syenitic and dioritic greenstones, which lie at the base of the Arenig or Lower Llandeilo strata, are largely quarried in North Wales for paving setts, curbstones, road-metal, and to some extent for building stones. Extensive quarries in these rocks are worked on Penmaenmawr, and near Portmadoc in Carnarvonshire; while down the north coast of the promontary of Lleyn, in the same county, rocks more nearly approaching a syenitic granite are largely quarried for the same purpose.

The costs of quarrying and forming a ton of paving setts, at one of these quarries when in full work, may be taken as follows:—

	Per ton of Setts. s.	d.
Quarrying, including removal of top rock . .	2	6
Sett making (average price)	9	0
Royalty	0	2
Powder and fuze	0	5
Management in and out of quarry . . .	1	9
Trammers and labourers	1	3
Loading, smiths, and contingencies . . .	0	10
	15	11

Basalt, consisting of augite, olivine, and felspar, is largely quarried for setts and road materials on the Clee Hills, Shropshire.

For particulars as to the position, varieties, and costs of quarrying the slate rocks of Wales, I will refer the reader to my work on that subject.[1] In a very few places portions of these are quarried for building stones, the most notable example at the present time being the extensive quarry recently opened out by myself and son in connection with the works of the Liverpool Corporation Water Supply, near Llanwddyn, Montgomeryshire. Here the beds are about three feet thick. The stone is too

[1] *Slate and Slate Quarrying*, by D. C. Davies, F.G.S. Crosby Lockwood & Co., 7, Stationers' Hall Court, London. Second Edition, 1881.

silicious for the cleavage to be perfect, although the lines are seen very distinctly. The rock crystallises in large masses in a rhomboidal form, the strike being N.E. to S.W., and the cracks or joints of shrinkage running to the east. The stone also cuts well along the line of its strike, answering to the pillari or pleri of the slate beds. In some of the beds where there is a little lime there is a tendency in the large blocks to break in five and six-sided columns. It is thus interesting to trace the same tendency to crystallise in particular forms, from the tiny crystal lining the side of a cavity, to the arrangement of rock masses on a large scale. This rock is exceedingly tough and hard, and capable of resisting great pressure. The general composition of the rock is silica 60, alumina 30, potash, soda, lime, and iron 10·0. A tunnel for the outlet of the water is now being driven through these beds at *8l.* 5*s.* yer yard, the length being about two miles; the cost of hand-boring is about 10*d.* per foot. Interstratified with these beds, there are all through North Wales beds of felspathic rock, very dense and compact. The cost of working in these is more than double that of ordinary slate rocks.

Of the same age probably as these felspathic and porphyritic bands or beds that over so large an area are interstratified with the Lower Silurian of North Wales, are some of the porphyries and serpentines of Cornwall, which furnish very beautiful building stones for in-door work.

Higher up in the geological series are the sandstones of the Coal-measures and of the New Red Sandstone, composed chiefly of silica, and being granular, and when first quarried somewhat soft and loose in texture, they are quarried with ease, the stones being often got out of the rock with wedges, and afterwards reduced to the desired size by the same method.

LIME AND LIMESTONES.

Calcium.—The metallic base of lime is the simple element calcium, which was made out, and its relative position to other metals assigned, by Sir Humphrey Davy. The name is derived from that given to lime (kalk) in several languages. Limestone consists of carbonate of lime, allied more or less with silicious, aluminous, and other matters. Pure carbonate of lime is seen in cracks and cavities in limestone masses, and in many stalactites and stalagmites occurring in large caverns in the same. It is composed of lime 56·0, and carbonic acid 44·0, the latter being made up of carbon 27·65, and oxygen 72·35. It occurs in masses chiefly white and light coloured, but ranging in some instances to black. It crystallises into various forms, of which two examples are given in figs. 5 and 6.

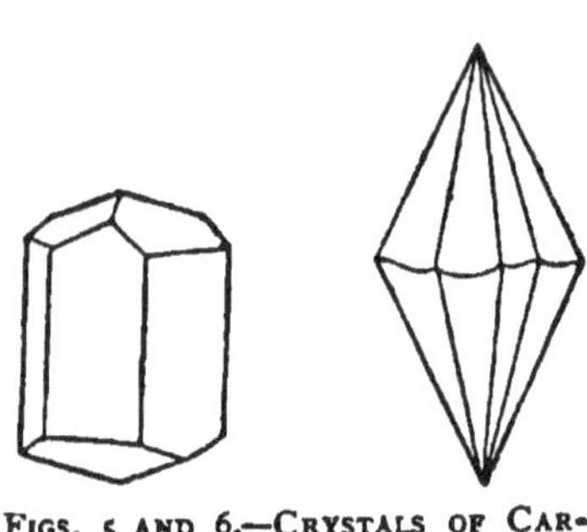

FIGS. 5 AND 6.—CRYSTALS OF CARBONATE OF LIME.

The crystals are of various colours, white, grey, yellow, and red, according as they contain metallic oxides or other matters. H. = 3; gr. 2·5 to 2·8.

Among the varieties of carbonate of lime are the following:—

1. *Argentine*, containing a little silica, with wavy laminæ, and of a white shining appearance.

2. *Calcareous Tufa*, a porus or cellular kind, formed in the hollows and in the vicinity of limestone strata, from water flowing over the latter, and becoming charged with carbonate of lime. Rock-milk is the name for it before it becomes consolidated.

3. *Chalk*, forming large masses of strata, soft, and rather earthy.

4. *Iceland Spar*, from Iceland, and famed for its double refracting property. In transparent crystals.

5. *Stalactite, Stalagmite.* Deposits like tufa formed in caverns and showing frequently rings or layers coloured by other mineral matter.

There are also the rock masses of limestone, that occur in almost every geological formation, from the oldest to the youngest.

Crystalline limestones occur in the Laurentian rocks of

Canada, as shewn in fig. 12, Chapter VII., and form the depositories of apatite.

There are the bands of limestone, Llandeilo and Bala, which occur in Great Britain, in the Lower or Cambro-Silurian strata, and in other counties in strata of similar age.

In the Upper Silurian there are the Wenlock and Dudley limestones, famous for the beautiful forms of sea organisms they contain.

In the Middle Devonian there are limestones which are seen in this country, but which are developed to a greater extent on both sides of the Rhine, in Germany.

This brings us up to the great mass of the Carboniferous Limestone, so well developed in Great Britain.

A belt of varying thickness skirts the north side of the South Wales coal-field. A similar belt bounds the west side of the North Wales coal-field, on the borders of England and Wales. This dips under the great red sandstone plain of Cheshire, and re-appears in the Derbyshire hills, whence it may be followed northwards through the counties of York, Durham and Northumberland, forming the great backbone of the country, and the depository of lead and zinc ores. Similar bands or belts encircle the Scottish coal-fields.

Fine marble and building stones are obtained from this group in Anglesea, while those of Derbyshire, with their abundant fossil remains, are well known. The same is true of the carboniferous limestone in other countries.

It is also very extensively quarried for agricultural use, being burnt in kilns and spread upon the land, where it helps to dissolve other minerals, and enables the plant to assimilate them. The light coloured and purer beds are also largely quarried for fluxing stone used in the smelting of iron, and also in the manufacture of glass.

The following particulars relative to the quarrying and burning of lime may be interesting and useful. They relate to the North Wales lime region.

Price paid to men for getting stone, and loading it in waggons, from 7*d.* to 8½*d.* per ton, the men finding their own powder, but drills and tools found, sharpened, and repaired by the owners. This is after the rock is stripped of its loose covering.

The contractors for the getting of the stone sometimes sublet the drilling to men whom they pay at the rate of 2*s.* 6*d.* to 3*s.* 6*d.* per 100 inches. No charging, tamping, or firing, is included in this price, and the men are found with tools. It will thus be seen that drilling in limestone is easier than drilling in granitic, gneissic, or dense slaty rocks. The men earn from 3*s.* to 4*s.* per day.

1 ton 15 cwt. of limestone makes one ton of lime, the rest, carbonic acid, sulphur, &c., being driven off in the process of burning.

1 ton of South Wales coal, from the district of Swansea and Neath, a quality between the anthracitic coal of the west and the bituminous coal of the east of that coal-field, will burn four to five tons of limestone, but the free-burning coals of North Wales and Lancashire will only burn from two and a half to three tons.

These prices do not apply to limestone quarried for building purposes, when more attention has to be paid to both size and shape.

Rising higher in the geological scale, we have the magnesian limestones of Nottingham, York, and Durham, which form hard and durable building stones.

In the Lias we have in England extensive quarries in its layers of limestone in central England, in the district between Birmingham and Oxford and Birmingham and Bletchley, as well as near Barrow-on-Soar, in Leicestershire. These limestones are valuable for their cement-making properties.

In the Oolites we have the Portland limestone and the calcareous sandstone, near the base of the series known as Bath stone, so valued in the west of England for architectural purposes; the Kenton and Ancaster stone, also from the great Oolite, used in ecclesiastical structures in the east of England. Of this age is the fine building stone of Caen, a calcareous freestone, and which is also found over large areas of France.

Of a similar age are the beautiful marbles of Carrara, in Italy. The stone has an extensive range in the Apennines, but the best quarries are those of the valleys in the neighbourhood of Carrara.

The best kinds are pure white and crystalline, but the general colour is a light blue or white, with bluish veins. The stones are quarried up to a large size, ten to fourteen feet in length, but large quantities are wasted owing to the want of mechanical appliances in the quarries, the stone being allowed to fall and tumble a long distance down a rugged rock face on heaps of débris. The quarries are supposed to have been worked since the first century of the Christian era.[1]

There are a few hard bands near the top of the chalk which have been utilized for building, but the mass of this formation is too soft for architectural use. Portions of the beds have, however, an agricultural value as fertilizers, and there are other uses to which the chalk beds have been applied.

The limestones higher in the series of strata are, as we shall see, the sources whence other valuable products are obtained, if they are not of so compact and durable a nature as some of the older limestones for building purposes.

GLUCINA.

Glucina is composed of two parts of glucinum, with three parts of oxygen. The metal glucinum is obtained with difficulty from its chloride. The process is much the same as that of aluminium. The name comes from γλυκος, sweet, on account of the sweet taste of its oxide, glucina.

The metal is not oxidizable by air or water at the usual temperature, but it takes fire in oxygen at a red heat, and burns with a vivid light.

Glucina is of rather rare occurrence, but, as we have seen, it enters to the extent of 15 per cent. into the composition of

[1] For a full description of the marble and other quarries of Great Britain and foreign countries, the reader is referred to *A Treatise on the Building and Ornamental Stones of Great Britain and Foreign Countries*, by Edward Hull, M.A., F.R.S. 1872.

the emerald, and common emeralds are found in thick crystals several feet in length in felspar quarries, in the parishes of Kisko, Roslagen, and Tammela, in Finland, from which the earth and its metal may be obtained. The minerals containing this earth have a specific gravity of 2·7 to 3·75, and with one exception—leucophane—they are harder than quartz, and are scarcely fusible before the blow-pipe.

ZIRCONIA.

Seems to be composed of about two parts of zirconium to three of oxygen. When obtained from its earth, zirconium appears as a powder, which may be compressed into scales resembling graphite, and burnished assumes the lustre of iron.

Zirconia occurs in four and eight sided prisms, and also in a granular state. Its colour ranges from white to grey, yellow, red, brownish red and brown. H. 7·5 and gr. 4·0 to 4·8; opaque to transparent. Common zircon occurs near Brevik, in the South of Norway, and crystals of the mineral occur at various places in the United States of America. Among the varieties of zircon we must notice the hyacinth, which is a transparent red and orange-coloured variety. Jargon or jargoon is a nearly colourless variety, with a smoky tinge. Its composition is zircon 65, silica 31, oxide of iron 2. It occurs in small crystals of four or eight sided prisms, with terminal pyramids. In lustre it approaches most nearly the diamond. The hyacinth and jargoon are found with other precious stones in the sands of Ceylon, already described. Crystals as large as walnuts come from Siberia, and fine specimens are brought from Greenland. Zircon is found in crystals at various places in North Carolina, Vermont, and New York. It belongs to the older granitic and gneissic rocks, lavas, and crystalline limestones.

THORIA.

The simple element thorium is the metallic base of thoria. It bears a general resemblance to aluminium. It burns in oxygen with an extraordinary degree of brilliancy. It was discovered in the year 1824 by Berzelius, in a black lava-like mineral, since named thorite after the Scandinavian god Thor, on the west coast of Norway. Thorite is black in colour, and heavier than the other earths, its gr. being 4·6 to 5·3.

CHEMICAL COMPOSITION OF VARIOUS LIMESTONES.

	Carboniferous Limestone from Porthywaen, near Oswestry, Greybeds.	Carboniferous Limestone from Minera, near Wrexham.	Dolomitic and Magnesian Limestone. General Composition.	White Magnesian Limestone, Nottinghamshire.	Portland Limestone Oolite.
Carbonate of lime . .	86·20	97·520	45·6	26·50	95·16
Carbonate of magnesia .	3·23	0·493	44·4	17·98	1·20
Silica	2·32			51·40	
Alumina		0·725	}	1·32	0·50
Oxide of iron	1·28				
Organic matter and water of combination	4·66	0·878			
Alkalies	1·59				
Water and goss . . .	72	0·384		2·80	1·94

CHAPTER III.

CLAYS.

Clays, how derived and formed—Decomposition of Basalt, of Rocks in Iceland—Brick Clays of the Drift—Analyses—Clays of the Tertiary Strata, Bovey Tracey—China Stone and China Clay of Cornwall and Devon—Comparison with other China Clays—Kimmeridge Clay—Permian Marls and Clays—Clays of the Coal-measures—Analyses—Manufactures of, in North of England, North Wales, Shropshire—Colouring of Clays and their Products.

CLAY may be described generally as decomposed silica with alumina (the former mineral usually predominating); and the clays of all ages have been derived ultimately from the wearing down and decomposition of rocks already described as containing these two minerals in their many varieties. But clays so derived do not contain all the constituents of their parent rocks. Some of these during the process of the chemical change we call decomposition have been removed, and by the residue an amount of water ranging from 10 to 20 per cent. of the mass has been taken up. Some interesting experiments illustrative of this were made by Ebelmen on the basalt of Auvergne, both in its unaltered and decomposed state. He found that in the process of decomposition there had been removed of the original constituents two-thirds of the silica, nine-tenths of the iron, which was protoxide in the original and peroxide in the decomposed rock, one-half the lime, $\frac{88}{100}$ths of the magnesia, and five-sixths of the potash and soda, together equal to 43 per cent. of the basalt. The alumina alone remained undiminished, a quantity of water having become united with that and with the other residual elements of the rock.

It has been found that the decomposition of the silicates of rocks (whether igneous or sedimentary) has been effected by means of oxygen and carbonic acid, the latter decomposing the silicates, and the former changing the protoxide into peroxide. Silica is readily soluble in water containing alkaline carbonates. It is also soluble in pure water, and in water charged with carbonic acid. Lime and magnesia are also soluble in water charged with carbonic acid.

The water flowing through the crevices and interstices of the earth's crust is so charged, and as in some places the waters of the sea and lakes are now so charged, so through past ages there have been times and places where the waters of seas and lakes have been so charged with carbonic acid. Hence by prolonged action the rocks in contact with these waters have had their silicates dissolved, the latter entering into new combinations, sometimes, indeed, as nearly pure silica, but usually with the alumina set free from the dissolved rocks, and with portions of the water forming clays. An illustration of the active and rapid decomposition of rocks occurs in the volcanic sulphur districts of Iceland, where various gases on their eruption from the earth decompose the surrounding rock (palagonite, a hard, reddish-white vitreous rock), and changes it into masses of clay of various colours according to the amount of iron and other minerals it contains.

Beds and irregular deposits of clay occur in all the groups of strata from the most recent down to the base of the Coal-measures. Thin layers and pockets occur in the underlying millstone grit. The softer layers of the carboniferous Limestone and of the underlying Devonian partake more of a shaly and marly nature. When we reach the Silurian and Cambrian, apart from clayey matter filling cracks and interstices with some thin beds in the Upper Silurian, it would seem as if whatever deposits of clay had been found in those early ages had again by heat and pressure become hardened into rock. Possibly the purer and finer kinds of felspathic rocks, such as those which in this country and in Norway and Sweden are quarried for porcelainitic purposes, represent these ancient clay beds.

Let us now notice the nature and uses of the various deposits of clay in stratigraphical order, beginning with the newest and uppermost.[1]

Starting with the north-west side of Britain, we find a thick deposit of gravel, sand, and clay covering the various rocks. Lying over nearly the whole of these deposits there is spread a thick covering of clay with boulders, known as the upper boulder clay. In hollows lying on the surface of this clay there are deposits of finer clay which has been washed out of the boulder clay, and has been redeposited comparatively free from stones in the hollows which at one time contained deep still water. It is of this clay, which is of a reddish yellow colour, that the surface bricks of the Welsh border counties are made, as well as many of those manufactured in Lancashire. These burn of a bright red colour. Many examples of these deposits may be seen in the brickyards between Chester and Oswestry.[2]

Where these clays spread over the Triassic strata, as near Shrewsbury, they become much redder in colour. The following analysis will show the general composition of these superficial clays:—

	1	2	3
Silica	66·68	49·38	57·83
Alumina	26·08	34·26	20·55
Oxide of iron	1·26	7·74	7·75
Oxide of manganese			traces
Lime	0·84	1·48	1·68
Magnesia	trace	5·14	0·97
Potash			3·87
Soda			0·56

[1] The reader will find much important and interesting information in 'The Source of the Materials composing the White Clays of the Tertiaries,' by George Maw, Esq., F G.S. *Quarterly Journal Geol. Society*, vol. xxiii. p. 387, *et seq*. Also *Catalogue of Specimens of the Clays and Plastic Strata of Great Britain, exhibited in the Museum of Practical Geology, London*, by the same author.

[2] See 'The Drift of the North Wales Border,' by D. C. Davies, F.G.S. *Proceedings, Geologists' Association*, vol. iv.

	1	2	3
Carbonic acid			0·90
Phosphoric acid			traces
Organic matter			4·39
Water	5·14	1·94	2·13

In the valleys of Wales a dark blue clay underlies the sand and gravel deposits, and it is probably derived by washing from the blue and grey slaty and shaly rocks that prevail in that country. A similar deposit is described by Mr. Maw as occurring underneath the alluvium of the Severn Valley, near Bridgenorth.

Underneath all the driftal deposits there are in North Wales and on the borders, nestling in hollows and troughs in the carboniferous limestone and millstone grit, numerous deposits of fine white clay which have evidently been dissolved out of the adjacent rocks, the lime having been carried off and in some cases deposited as tufa in the immediate neighbourhood. The white clays of Nant-y-Garmer, near Llandudno, and of Halkin and Mold Mountains, occurring in pockets and irregular masses, are examples of these. The red and dark marls and shales of the limestones have also contributed to these clays, examples occurring in the red clay of Nant-y-Garmer, Llandudno, and in the dark clay of Llanferris, near Mold.

The China or Porcelain Clays of Cornwall.—Although derived from rocks of a different nature and age, these deposits may be grouped from their similarity of origin and position with those just referred to. The china clay trade of Cornwall is of comparatively recent date. It appears to have originated with Mr. William Cookworthy, who opened a pottery at Plymouth in the year 1733, at which he worked some of the clays of Devon. In 1755 he found a stone near St. Stephen's, the present china stone, which he used for forming a glaze upon the porcelain. In 1774 Cookworthy sold his patent to Richard Champion, a Bristol merchant, and the works were removed to that city. Soon afterwards the works were transferred to Tunstall, in Staffordshire, where the business grew and flourished, the clays and stones of Cornwall thus finding their way to the Staffordshire potteries. The largest deposits of china clay in Cornwall are found in the neighbourhood of St. Austell and St. Stephen's. There are also other deposits in the eastern part of the county, at Blisland and St. Breward, near Bodmin, and also near Helston on the west. From the St. Stephen's and St. Austell district there were exported to the potteries in the year 1809, 1,757 tons; in 1810, 1,888 tons; in 1811, 2,086½ tons; and in 1812, 1,252 tons. Besides the exports to Staffordshire there were sent to the china manufactories of Worcestershire, from March, 1816, to March, 1817, 1,775 tons. By the year 1826 the trade had grown so that a total of 7,538 tons were shipped that year from Cornwall, the points of production having also extended to other places, the district of St. Austell and St. Stephen's, however, supplying 7,090 tons of the amount. In 1838 the production of china clay in Cornwall was estimated at 7,600 tons per annum. The production in the year 1880 was 278,572 tons, and in 1881 241,658 tons. In addition to this the production of china stone, which in 1838 was estimated at 5,000 tons, was in 1880 34,870 tons, and in 1881 30,479 tons. Besides these quantities there were also raised in Devonshire, from a different deposit for the most part, in 1880 25,370 tons, and in 1881 39,067 tons. The price of the china stone delivered free on board ship is from 30*s.* to 40*s.* per ton, that of the china clay about 25*s.* per ton. The royalty payable to the landowner on both is from 2*s.* 6*d.* to 4*s.* per ton, a rather heavy royalty. The poor-rates amount to 8*d.* per ton.

The china clay has been derived from the decomposition and wearing down of the granitic rocks in the neighbourhood of the deposits, especially of those parts which yield most readily to those atmospheric and aqueous influences referred to at the commencement of this chapter. The materials have been redeposited in adjacent hollows and flats, and in some cases where the rock has been crossed and reticulated by veins or lodes, the decomposed material remains in its original position. An example of this occurs at the Carclaze pit, situated about two miles north-east of the town of St. Austell.

It is one of the largest open excavations for clay in Cornwall, being about 14 acres in extent. Tin has been mined here as in an open *stockwerk* for several centuries. The tin was found in numerous veins from 2 in. to 2 ft. wide, that traversed in every direction a decomposed granite of a whitish colour. The quarry is now worked almost exclusively for clay.

When quarried the clay is separated from the coarser materials with which it is associated by washing. It is placed on either a natural or artificial gently inclined plane. A stream of water falls upon it from the height of a few feet, which gradually washes it away. The larger fragments are caught upon gratings. There are dams placed at distances of 20 ft. or so apart, which also intercept the heavier materials. The liquid then flows through a series of tanks, until at last only the very fine clay remains. This bears the proportion of 1 ton to every 8 tons quarried. It is cut into lumps of convenient size, is dried, formerly without the application of artificial heat, but now this is frequently used, and when it is sufficiently hard to bear removal the clay is ready for sale. The wages of the men averages 2*s.* 6*d.* per day of 7½ hours. There are about 100 clay works in Cornwall, employing about 1,600 workmen.

The composition of St. Stephen's china clay is given as, silica 39·55, alumina 38·05, water 12·50, magnesia a trace, with an insoluble remainder of 8·70, probably chiefly silica. An analysis of china stone from the parish of St. Roche shows its composition to be silica 63·17, alumina 20·89, peroxide of iron 0·14, lime 0·90, magnesia 0·21, potash 11·48, soda 3·11, total 94·90. It will be seen that in the clays the potash and soda, with portions of the silica, have been removed.

It may be interesting here to compare the composition of the china clay of Cornwall with other clays used in the manufacture of porcelain.

Original kaolin[1] of China.	Silica 76, alumina 17, potash and soda 6 (the water previously removed).

[1] Kaolin, a corruption of *Kauling*, a high ridge—the name of a hill near Jauchau Fu, where the mineral is obtained.

Original kaolin of Japan.	Same as above, only with less alkalies.
Kaolin or porcelain clay from Schneeberg.	Silica 43·6, alumina 37·7, peroxide of iron 1·5, water 12·6.
Kaolin from Seilitz, Saxony, after water was expelled.	Silica 54, alumina 44, potash trace.
Kaolin from Morl, near Berlin.	Silica 71·4, alumina 26·0, lime and potash traces.
„ St. Yriex, France.	Silica 46·8, alumina 37, water, 13, potash 2·5.
„ Dartmoor, Devon.	Nearly identical with the foregoing from St. Yriex.
To the foregoing I may add the composition of the fine red clay, of which the North American Indians make their pipes, and named after the adventurous traveller, Mr. Catlin, Catlinite.	Silica 48·2, alumina 28·2, magnesia 6 to 7, peroxide of iron 5, carbonate of lime 2·6, oxide of manganese 0·6, water 8·4.

Where felspar can be found of sufficient purity it is quarried for pottery or porcelain manufactures. This is done as already stated in Cornwall, and considerable quantities are imported into this country from Sweden and Norway. In Ireland, too, the beautiful porcelain of Fermanagh is obtained from the red felspar found in the vicinity. When burnt this felspar loses its colour and becomes white, and the metallic iron, from whose presence the colour was derived, is separated from the powdered felspar, when it is mixed with water, by means of magnets.

Among the clays of the Tertiary strata especial reference may be made to those of Bovey Tracey, referred to in Chapter XII., as associated with the bituminous deposit of the same place, and which form an important source of the clays exported from Devonshire for pottery purposes.

These clays are divided into several kinds according to their colour, texture, and the chief purpose for which they are employed. There are the 'best pipe clay,' the 'cutty clay,' the 'household clay,' which is used for whitening stone steps and pavement; the 'stoneware clay,' employed in the manufacture of stoneware; the 'alum makers' clay,' used for pottery purposes and the manufacture of alum; the 'drain-pipe clay,' used in the manufacture of drain-pipes and for other common purposes; it is stained with iron; the 'blue ball clay,' which burns to a pale colour, and is extensively used for earthenware purposes in the potteries, and which shows on analysis the following composition:—

Silica	47·0
Alumina	48·0
Oxide of iron	1·5
Magnesia	2·0
Water and waste	1·5
	100·0

The 'black ball clay,' the dark colour of which is probably due to the presence of carbonaceous matter; the 'brown ball clay,' employed for pottery purposes (composition on analysis shows two-thirds silica, one-third alumina, with traces of magnesia, oxide of iron, and carbonaceous matter); the 'black carbonaceous clay.' This clay in its natural state contains 13 per cent. of carbon, and the coarser kinds 33 per cent. It burns to extreme whiteness, which is believed to be due to the reduction of the sesquioxide of iron it contains in the kiln by reaction with carbonaceous matter.

Lower down still in the Tertiary strata are the clays worked near Wareham, Dorsetshire. These consist of a 'red plastic clay,' which is used to a small extent in the manufacture of encaustic tiles; 'white clay,' which fires of a light cream-colour, and is employed for various pottery purposes; '7 clay,' also employed for pottery uses, and is remarkable, with the following, occurring at the same place, for its great contraction in the kiln; 'black clay,' coloured with carbonaceous matter which burns out in the kiln; 'blue clay,' containing about 60 per cent. of silica, 34 of alumina, 2 of potash, with small amounts of oxide of iron and water of combination; 'yellow clay' of a bright yellow colour, due to the sesquioxide of iron being hydrous. It is associated with the blood-red plastic clay first named. These two clays are used in the manufacture of encaustic tiles and are worth from 7*s.* to 10*s.* per ton free on board at Poole harbour.

There are various beds of clay in 'the London clay' worked near Harwich, Bognor, Arundel, and others interstratified with the cretaceous Wealden and Purbeck strata which are locally worked, among which may be named the 'fullers'-earth' of the Sandgate beds of the Lower Greensand. This occurs between Red Hill and Nutfield, Surrey. It also occurs near Maidstone in Kent, Woburn in Bedfordshire, and near Maxton, in Scotland. The fullers'-earth of Surrey is composed of silica 53, alumina 10, iron peroxide 9·75, magnesia 1·25, lime 0·5, and water 24·0. In the Oolitic series among other clays there is the well-known Kimmeridge clay, used for making coarse pottery and bricks, and which, in addition to silica and alumina, contains protoxide of iron 2·08, sesquioxide of iron 4·32, bisulphide of iron 1·42, carbonate of lime 4·28, and sulphate of lime 5·34, with crystals of selenite distributed throughout the mass of the strata.

Passing by the Oxford and other clays of the Oolite and those of the Liassic, Rhaetic, and Triassic strata, some of which are of local importance, we come to the red marls of the Permian, which form the base of important and extensive industries in Denbighshire, Shropshire, and Warwickshire. These marls form the middle division of the Permian strata. They are, in Denbighshire and the contiguous parts of Shropshire, about one hundred and fifty yards in thickness. They are of a deep red colour, but contain nests and irregular patches of a light green, greenish grey, and buff colour. These patches are carefully separated from the mass in quarrying, especially for the finer kinds of work. The quarries are deep open excavations, from which the clay is usually drawn up an inclined plane to the highest point of the works, whence, after it has been crushed, ground, and duly mixed, it is distributed over the works for the various processes it has to pass through. The red portions, containing most oxide of iron, burn of a deep red colour, although the presence of the alkaline earths in clays containing most iron will cause the product to be of a lighter colour. There is scarcely any limit to the uses to which these clays are put—bricks, flooring and roofing tiles, drain-pipes, terra-cotta work, cornices and architectural mouldings and decorations of all sorts. A visit to the works of Mr. J. C. Edwards, Penybont, near Ruabon, or to those of Messrs. Maw, near Brosely, will show the extent to which, by means of chemical knowledge and long experience, the clays are made use of for many purposes. The works in connection with the Hockley Hall Collieries, near Tamworth, are also well worth a visit, presenting as they do the example of works skilfully laid out for producing the largest results at the smallest cost.

Besides 80 per cent. of silica and alumina, the red portions of these marls contain:—

Water of combination	2·79
Sesquioxide of iron	3·23
Protoxide of iron	1·35
Bisulphide of iron	0·02
Carbonate of lime	4·15
Sulphate of lime	0·17
Alumina	3·95
Magnesia	2·17
Alkalies and loss	1·22

In the light-coloured patches the proportion of iron is:—

Sesquioxide	1·81
Protoxide, soluble	0·60
Protoxide, insoluble	0·88
	3·29

Some examples of the red portions contain from 7·54 to 8·64 of sesquioxide of iron, the lighter colour of the green and grey patches being due to the smaller quantity of iron in their composition.

The Clays of the Coal-measures.—Underneath most of the beds of coal in the British, as well as other coal-fields, there is a bed of clay varying from 1 to 18 feet in thickness. At the base of the Permian the red colour for the most part ceases, and is replaced by clunch, or clay of a blue, yellow, or pale buff colour. The light-coloured clays, from the quantity of silica they contain, as well as from their freedom from impurities, have great fire-resisting properties, and are known as fire-clays. During the last and present centuries a large industry has grown up in the use of the clays of the Coal-measures for various purposes, more especially in connection with the north of England, Flintshire, Denbighshire, Shropshire, and Staffordshire coal-fields.

In the north of England the manufacture of earthenware was started about the year 1830, and that of fire-bricks and fire-clay goods generally, first started on the Tyne, about a hundred years ago. The clays under different coal seams possessing different properties, and all of them being capable of being worked from the same mine—if deemed desirable—gives great facilities for the manufacture of widely different goods at clay works established in connection with a colliery. Mr. Joseph Cowen, M.P., gives[1] the following analyses of samples of fire-clay from seven beds underlying coal in the neighbourhood of Newcastle-upon-Tyne :—

	1.	2.	3.	4.	5.	6.	7.
Silica . . .	51·10	47·55	48·55	51·11	71·28	83·29	69·25
Alumina . .	31·35	29·50	30·25	30·40	17·75	8·10	17·90
Oxide of iron .	4·63	9·13	4·06	4·91 }	2·43 }	1·88 }	2·97
Lime . . .	1·46	1·34	1·66	1·76 }			
Magnesia . .	1·54	0·71	1·91	trace	2·30 }	2·99 }	1·30
Water and organic matter	10·47	12·01	10·67	12·29	6·94	3·64	7·58

The fire-clays—those containing the most silica—are made into bricks for lining furnaces, sanitary tubes, gas-retorts, architectural ornaments, and many other articles which the modern requirements of science, sanitation, and manufactures demand.

Passing south-west to the North Wales and Shropshire coal-field, the following remarks from a paper of mine published a

[1] *The Industrial Resources of the Tyne, Wear, and Tees.* Newcastle, 1864.

few years ago,[1] may serve to describe both the growth of the industry and the purposes for which the various clays are used.

The following is a very general section of the order and position of the principal coal-seams of North Wales and the contiguous part of Shropshire, and its perusal will enable the reader to understand the subsequent allusions to the different clays and works of the district :—

		Strata	Yds.	Yds.
	Permian.	*Dark Red Sandstones.*		
Upper		Saint Martin's Coal-measures		90
Middle		Red marls		112
Lower		Grey and green rocks and conglomerate		130
Upper . .	Coal-measures.	Upper coal-measures		90
Middle . .		Middle sandstones, shales, and coals, including the freestone known as the Cefn, Minera, or Hollin sandstone		140
Lower .		*Cefn Coal.*		
		Clays, shales, and thin coals	100	
		Dirty or Drowsal Coal.		
		Strata as under Cefn coal	60	
		Quaker Coal.		
		Strata as before	25	
		Main Coal.		
		Strata as before	14	
		Upper Yard Coal.		
		Strata with coals	70	
		Lower Yard Coal.		
		Strata	18	
		Wall and Bench Coal.		
		Strata	35	
		Llwynenion or Half Yard Coal.		
		Strata	35	
		Chwarele Coal.		
		Strata	25	
		Lowest Coal.		292
		Add thickness of coals		12
				866
Millstone grit.		Sandstones.		

The clays of the 'Coal-measures' are usually denominated

[1] 'The Fire-clays and Fire-clay Productions of North Wales,' *British Architect*, Jan. 21st and 28th, 1877.

'fire-clays.' Their power of resisting the action of fire is, however, very variable. The best adapted for this purpose are those that contain the largest proportion of silica, and that are free from any admixture of the oxides or sulphides of iron. Even the best of them have their fire-resisting quality increased by the addition of sand from the beds of the millstone grit; indeed, the bricks made from the 'gannister' beds of that series are the most largely used in the manufacture of iron and steel. Mr. Thomas Barnes, of the Quinta, and Mr. J. C. Edwards, of Trefynant, have both of them made some interesting experiments in the admixture of these sandstones with the clays worked by them.

We now turn to the notice of the principal brick and tile works of the district, and we find that among the earliest works of the kind were those of the late Mr. Howell, known as 'The Pottery,' at Trefonen, south-west of Oswestry. The writer remembers these in active work forty years ago, and at that date the proprietor had discovered the adaptability of the clays to the manufacture of pipes, tile-crests, chimney-tops, and many of the other purposes for which, since then, they have been more extensively used.

The chief clay worked at Trefonen was the one underlying the four-foot coal of the locality, and which corresponds to the Quaker coal of the section. These works were closed about twenty-five years ago, when the bulk of the trade was removed to the works of the late Messrs. Croxon, at Sweeney, south of Oswestry. The manufacture of fire-bricks, tiles, and drain-pipes has been largely carried on at these works until now, and recently, the concern having changed owners, the present company—the Oswestry Coal and Brick Company—have made extensive additions to the works, to meet the requirements of an increased business.

Two clays chiefly are used here, the one under the 'four-foot coal' as at Trefonen, and that under the 'black shale coal,' a coal answering to the 'Drowsal coal' of the section. It was from an old pit-heap composed largely of these clays that the bricks used in the construction of the Oswestry sewage works were made. The clays themselves yield a nearly white brick, that looks well in a building, although, perhaps, the absence of warmth of colour may, to some tastes, be a defect. The owners have, however, on their property a thick deposit of the 'red marls,' and by a judicious mixing of these with the white clays good results may be obtained.

A very important work, full of good machinery and connected with a railway, was some years ago carried on by Mr. Thomas Savon at Coedygo, midway between the two former works. It is now closed, and every vestige of it removed, the owner of the freehold, it is said, objecting to such works as spoiling the rurality of the neighbourhood.

Six miles to the north of Sweeney are the brick-works of the Quinta Colliery and Brick Company. The clay of the 'yard coal' was worked here for some years; but latterly the under clay of a lower coal, probably that of the 'lower yard,' has been substituted. The former clay contained an appreciable quantity of oxide of iron, which occasionally gave a reddish tinge to the bricks. An improvement has ensued with the working of the lower clay, and a nice sound brick, with rather more colour than those of Sweeney, is the result. The manufacture here is not confined to bricks, but includes pipes, tiles, &c.

Three miles farther north, and situated on a promontory formed by the windings of the river Dee, we reach the Penybont Works, belonging to Mr. J. C. Edwards, of Trevor. The clay used at these works is obtained in an open excavation from the 'red marls' of the Permian, near the top of the section. Following the colour of the clay, the bricks and other productions at these works are of a deep red colour. Perhaps a better red brick, smoother, more uniform in texture and colour, and more impervious to moisture, is not produced than the best pressed bricks from the clays of these works. In addition to bricks, the manufacture of blue paving-tiles is largely carried on, as well as that of roofing-tiles, for which, owing to the scarcity of roofing-slates, and the revived taste for red-tiled roofs, there is a good demand. Other articles, such as crests, pipes, flooring-tiles, &c., are also extensively made here.

About a mile and a half higher up the valley of the Dee are Mr. Edwards's older fire-clay works, at Trefynant, which may be briefly described as among the most complete works of the kind in North Wales. The clay used is that underlying the 'Llwynenion,' or 'Half-yard coal.' This clay is many feet in thickness. The underground workings are extensive, and the coal is left in the ground to form a roof for them.

The productions are of a pale buff colour, of great uniformity of appearance. Mr. Edwards has, from time to time during a long course of years, added the manufacture of chimney-tops, socket-pipes, junctions, and of all kinds of sanitary ware, to the glaze and finish of which he has given much attention.

Of a very similar kind, and deserving of much the same remarks, are the more recent works of Mr. Bowers and of Mr. Seacombe, nearer to the village of Ruabon.

These gentlemen, with Mr. Edwards, have hitherto been the chief producers of sewage, drainage, and sanitary appliances in North Wales. The bricks from all the works command a large sale.

A little farther to the west of the Trefynant Works we are at Garth Trevor and on the lower edge of the Coal-measures. Here, from the shaly and iron-stained clays and sandstones that lie between the Chwarele and lowest coal, Mr. Charles Mason produces large quantities of good sound and serviceable cherry-red bricks, which are well adapted for all ordinary building purposes. The clay of the 'Chwarele coal' occurs abundantly on this property. Its productions are of much the same character as those of the Llwynenion clay, and it is well adapted for ordinary pottery use. It has not, however, as yet been worked to any extent by Mr. Mason, as it has in the adjoining 'Chwarele Works,' of which little can be said just now, except that, through the vicissitudes of trade, operations were recently suspended.

Up the hill to the north, and situated about two miles to

the west of the village of Ruabon, are the more recently developed works of Messrs. Smith and Thomas, at Plas Ucha. This firm works the clays associated with the 'wall and bench' seams of coal, which here are adapted for use, and yield good bricks of a pale yellow colour that are rapidly finding their way into the market.

More recent still are the works at Plas-yn-wern, the property of Mr. G. H. Whalley, M.P., and which promise to be among the most extensive of the district. Various clays, from the 'Quaker' downwards, are intended to be used at these works.

At Ponkey, farther north, good hard pale red bricks are made from the clays in the refuse heaps of old coal-pits, and one or two works are in the early stages of growth.[1]

A stretch of six or seven miles brings us to the very edge of the Denbighshire coal-field, where it is separated from that of Flintshire by the limestone and grit hills of Hope and Caergwrle. Here are the Llay Hall Colliery and Brick Works, which are just being brought into shape, and where, in addition to the use of the clays from the lower coal-seams, the owners intend to work the series of clays and shales that lie in the middle series between the 'Cefn' or 'Minera sandstone,' and the 'Cefn' or 'stinking' coal.

The mention of these shales and clays, and a walk of a few miles across the dividing hills, brings us to the old and great brickmaking region of Buckley Mountain. The works on and about the 'mountain' are very numerous, and among the oldest are those belonging to Mr. Hancock and Mr. Catherall. Great open excavations meet us at almost every turn. These are dug in the shaly clays just alluded to as underlying the 'Cefn' or 'Minera sandstone,' which here is known as the 'Hollin coal rock,' or sandstone. Immediately underneath this rock is a thick deposit of rocky clay, which, mixed with a little of the purer clays, makes a good common brick. Next below is a series of blue and yellow clays, which yield good

[1] As these pages are passing through the press, most extensive terra-cotta and brick works are being erected near Ponkey.

bricks of a pale red colour. Lower down is a dark clay, which, when burnt, loses its dark carbonaceous colouring matter, and gives a nice pale-coloured fire-brick.

Underneath the whole of the above clays there is on the 'mountain' a bed of fine yellow clay, which has long been used for the common, though useful, kind of pottery ware for which the place is famous. A bed of the same character, and occupying just the same position, has also been worked for many years at Cefn, near Ruabon.

On the northern slope of Buckley Mountain, and lying between it and the river Dee, is the Aston Hall Colliery and Brick Works, on the property of the Right Hon. W. E. Gladstone, M.P. A beautiful rich-coloured yellowish brick is made here from the clays of 'wall and bench coal,' as at Plas Ucha, near Ruabon. These bricks may be seen to advantage in some of the best buildings at Rhyl and other towns on the north coast of Wales; they form a nice contrast to and in combination with the dark red bricks from the 'red marls' at Penybont.

Since the foregoing remarks were written, changes have taken place in the ownership of the smaller works; but Edwards's, Bowers's, and Seacombe's remain in the same hands. At the Sweeney Colliery the red clay referred to has since been opened upon and worked with considerable success.

The clays of the Coal-measures are hard, and are usually mined by cutting away the foot and boring in the top of the seam for blasting. The price paid to the getters varies from 1*s.* 8*d.* to 2*s.* 3*d.* per ton, according to the hardness and thickness of the seam. In most cases the newly-obtained clay is laid out conveniently upon the surface for weathering before it is used. The purer yellow and buff clays are much mixed in the district with the red clays of the Permian in the manufacture of encaustic tiles.

The clays most used in the South Shropshire coal-field are the following:—

A *red marl* occurring from 20 to 30 yards above the 'sulphur coal' at Broseley Green. This is about 5 ft. thick, and it is extensively employed at the Messrs. Maws works, in association with more refractory clays, in the manufacture of encaustic tiles. The red portion of the clay contains up to 8·64 of iron; it burns of a dull red colour. Its composition is as follows: silica $\left\{\begin{array}{l}\text{combined}\\\text{free}\end{array}\right\} \left\{\begin{array}{l}29{\cdot}71\\34{\cdot}35\end{array}\right\}$ 64·06, titanic acid 0·62, alumina 2·60, sesquioxide of iron 6·84, protoxide of iron 0·32, protoxide of manganese 0·09, lime 0·12, magnesia 0·04, potash 0·91, soda 0·44, water, with traces of organic matter, 5·85. The *brick clay*, occurring about the middle of the Coal-measures as a bed from 8 to 12 ft. thick, and is extensively used in the manufacture of the celebrated brown Broseley bricks. The *Pennystone Mount*, which forms the matrix of the Pennystone ironstone at Benthall, Broseley. It is 6 ft. thick, and has nodules of carbonate of iron in the upper part; it is grey in colour, and of a rich ochreous brown after burning. It is used in the manufacture of encaustic tiles. Two-foot coal fire-clay, worked at Benthall, Broseley. Colour, dark grey; burns of a pale buff or cream colour. It is used in the manufacture of fire-bricks, common 'yellow ware' pottery, and encaustic tiles.

'Ganie coal fire-clay,' in the lower Coal-measures, about 2 ft. thick, of a grey colour, and burns of a pale buff or cream colour. It is one of the most refractory clays of the Shropshire coal-field. It is employed in the manufacture of bricks and encaustic tiles.

The clays of the three coal-fields already described may be taken as representative of those obtained from the other British coal-fields, from the whole of which there was produced of clay in the year 1881, as returned to the inspectors, 1,896,907 tons.

With its wealth of mineral resources, America possesses an abundance of clays of various kinds, which are gradually being utilised. The same remark is true of the Coal-measures and other clays of various countries; but the foregoing description of the clays of Great Britain may be taken as representative of similar deposits in foreign countries.

The following remarks relative to the colouring of burnt

clays, by Mr. George Maw, F.G.S., given in the 'Catalogue of British Clays,' to which I am indebted for much information, are interesting and valuable, the more so since they are the result of much investigation and practical experience.

'The colour of burnt ferruginous clays is entirely due to the amount of iron present, irrespective of its previous state of combination, but subject to certain conditions in the general composition of the clay. The action of the kiln, with some exceptions referred to below, is uniform on nearly every state of combination in which the iron occurs, viz. to reduce it to anhydrous sesquioxide, associated as silicates in a more or less intimate state of combination with the other silicates developed in burning.

'Yellow clays coloured with hydrous sesquioxide, *e.g.* yellow ochre, and red clays coloured with anhydrous sesquioxide and the lower hydrates, merely lose their water of combination and become bright red bricks.

'Grey clays, containing finely divided pyrites or bisulphide of iron, are also converted by the kiln into bright reds, the sulphur being driven off, leaving the terra-cotta charged with the red anhydrous oxide.

'In clays charged with grey carbonates of iron the following reaction takes place. The carbonic acid is driven off as carbonic oxide, part of its oxygen peroxidizing the iron.

'Grey clays containing less than 1 or 1½ per cent. of iron change in the kiln to various shades of cream colour or buff, whilst those containing from 2 to 10 or 12 per cent. of iron produce in the kiln the bright red bodies used in the manufacture of terra-cotta, encaustic tiles, red building bricks, &c. There seems to be no essential difference, with the exception noticed below, in the colouring matter of the clays that burn buff and those that burn red in the kiln, the depth of colour depending merely on the amount of iron present, the buff shades graduating into the deeper shades of red.

'The brightest shades of red and buff are, however, produced with but a partial vitrification of the body. At a heat sufficient to insure its complete vitrification a further change of colour takes place. The bright buff shades are changed to neutral greys, and the reds to a slaty greyish black, which probably results from a partial reduction of the metallic colouring matter, and its more intimate combination with the other vitreous silicates produced at the higher temperatures. In clays containing a large proportion of carbonaceous matter the complete peroxidation and consequent colouring power of the iron seems to be arrested. In the black carbonaceous clay of Bovey Tracey, containing 13 per cent. of organic matter, the combustion of the carbon in contact with the ferruginous oxides seems wholly or partially to have reduced them to a metallic state or lower oxide having less colouring power than the sesquioxide, and a remarkable bleaching of the burnt clay has been the result. The presence of the alkaline earths in ferruginous clays, especially of lime and magnesia, has also a singular bleaching power in the kiln, arresting the development of the bright red colour. One of the red marls of the Permian, containing 6 per cent. of sesquioxide of iron and 35 per cent. of carbonate of lime, burned of a greyish buff instead of the rich red such a proportion of iron would otherwise have produced. From some experiments made by the writer it has been ascertained that as small a proportion as 5 per cent. of caustic magnesia mixed with a red clay entirely destroys its red colour in the kiln, probably from the production of a pale-coloured silicate of iron and the alkaline earth. A familiar example of this reaction occurs in the process of manufacturing yellow bricks in the neighbourhood of London, the colour of which is dependent on the admixture of ground chalk with the brick earth, the brick earth by itself burning of a red colour.'

Clays occur in various degrees of fineness, from the coarser-grained clays of the drift to the fine clays of the tertiary beds and the Coal-measures. All clays contract in burning, but it is found that coarse-grained clays containing most silica contract less in burning than fine smooth clays. The amount of contraction is due to the loss of water, the loss of carbonic acid, and the consumption of the carbonaceous matter contained in the clay. It is also affected by the presence of alkalies, which promote the complete vitrification and consequent drawing together of the silicious particles. Clays made up also of large and small particles contract less than those made up of grains more uniform in size, the smaller particles helping to fill up the interstices of the larger. The average amount of the contraction of clays in burning is from 6 to 7 per cent. of the original moulded size of the article.

PART II.

HALOID MINERALS.

(COMPOUNDS OF THE EARTHS AND ALKALIES), CHLORIDE OF SODIUM (COMMON SALT), NITRATE OF SODA, BORAX, BARYTA, GYPSUM, ALUM SHALE, PHOSPHATE OF LIME.

CHAPTER IV.

SODIUM, CHLORINE, CHLORIDE OF SODIUM (COMMON SALT).

Description of Sodium—Of Chlorine—Common Salt resulting from the Combination of the Two—The New Red Sandstone Plains of England—Stratigraphical Position of the Cheshire Salt Deposits—History of Salt-working in Cheshire and Worcestershire—Strata overlying the Cheshire Salt Beds—Analysis of Rock Salt—Of Brines—Details of Cheshire Salt-mining—Brine Springs of Worcestershire—Associated Strata—Detailed Section of—Characteristics of the Brine—Statistics—Brine Springs of Ashby Wolds—Salt Manufacture in the North of England—Manufacture of Sulphate of Soda—Discovery of a Bed of Rock Salt at Middlesborough—Detailed Section of Strata—Analysis of the Rock Salt—Second Boring near Fort Clarence with Results—Rock Salt Deposits of Carrickfergus, Ireland—Method of Working.

SODIUM.

SODIUM, the metallic base of soda, and one of the simple elements, is a soft white metal, with the appearance of silver. It was obtained by Sir Humphrey Davy by the voltaic decomposition of soda. It may be cut with a knife, and it yields to the pressure of the fingers. On exposure to the air it oxidizes spontaneously, and when nearly red-hot it takes fire and burns with a yellow flame. It passes into a liquid state at a temperature of 194°. It is largely diffused throughout nature, its salts being found in all animal fluids, and, as we have seen, there is an appreciable proportion in most of the older rocks.

CHLORINE.—CHLORIDE OF SODIUM.

Chlorine was discovered by Scheele in 1774, and it was considered to be of a compound nature till 1809, when Gay-Lussac and Thénard demonstrated that it should be considered a simple substance. Sir Humphrey Davy followed up the investigation of it shortly afterwards, and gave it its name from χλωρος, yellowish green, in allusion to the colour of its gas, which is also very dense, and with a strong, suffocating odour. Sodium takes fire in this gas, and combining with that element, common salt is the result. The same result has been produced by the union of the two elements in the sea, and the proportions in which they have combined with each other and with other substances will be seen in the following pages.

From the summit of the Minera Mountain, five miles west of Wrexham, North Wales, the spectator has around him one of the most extensive views in England. Eastwards there is the New Red Sandstone plain of Cheshire; to the south-west he may discern the openings through which this

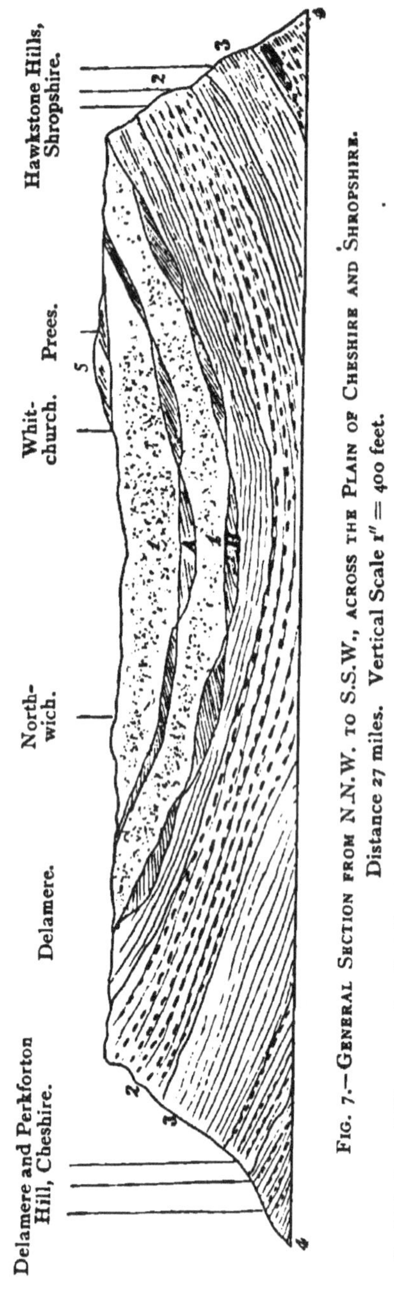

Fig. 7.—General Section from N.N.W. to S.S.W., across the Plain of Cheshire and Shropshire. Distance 27 miles. Vertical Scale 1″ = 400 feet.

1 Red Marls and Clays, see detailed Section . . . . Muschel Kalk wanting.
2 Flagstones and Marl } Keuper.
3 Freestones, Red and Light Colour } Waterstones }
4 Red Sandstones . . . . Bunter Top Beds.
5 Small Outlier of Lias.
A Upper Salt Bed.
B Lower Salt Bed.

geological formation extends down the plain of the Severn to Worcester and the Bristol Channel. With a little aid from his imagination he realises how the same red rocks to the south-east mantle around the southern termination of the Pennine Chain of carboniferous rocks, through the counties of Leicester and Nottingham, and form on the eastern side of the chain the Red Sandstone plains of York and Durham.

It is on the uppermost division of this strata, as shown in the section, Fig. 7, that the great salt deposits of Cheshire are found.

Before I describe the details of the way in which these deposits occur, I will give a brief *résumé* of the rise and progress of the industry. In doing this I am glad to avail myself of the help afforded by the information gathered some ten years ago by Mr. Joseph Dickinson, F.G.S., one of H.M. Inspectors of Mines, and which is embodied in his excellent 'Report on Land-slips in the Salt Districts.'

It is clear that the Romans, following in the wake of the original inhabitants, worked the salt springs, and among the earliest subsequent references to salt springs are those which relate to those of Droitwich, in Worcestershire. From these it appears that Kenulph, King of the Mercians, in the year 816, gave Hamilton and ten houses in Wich—the name that seems to have been given to places containing salt springs—to the church of Worcester; and about the year 906 Edwy, King of England, endowed the same church with Tepstone and five salt furnaces or scales.

Between the years 1084 and 1086 William the Conqueror caused among the other inquiries, the results of which are recorded in Domesday Book, one to be made relative to the Wichs and salt-houses—their names, by whom they had been held in the time of Edward the Confessor, the last hereditary Saxon king, and by whom they were held at the time the inquiry was made. In 1863 Mr. William Beaumont, after a painstaking examination of the German text and the numerous contractions, gave a zincograph of the original document, together with a translation, from which the following particulars were collected by Mr. Dickinson :—

'In Roeleau hundred the Earl Hugh holds Wyreham (Weaverham) in demesne. Earl Edward held it. A foreigner holds of the Earl. There were in Wych seven salt-houses belonging to this manor. One of these now renders salt to the Hall; the others are waste. The Earl himself holds Frotesham (Frodsham). There is in Wych half a salt-house to supply the Hall.

'In Dudestan hundred. Robert Fitz Hugh holds Beddesfield (Bettisfield, Flintshire) of Earl Hugh; Earl Edwin held it. The same Robert holds Burwardestone; Earl Edwin held it. There is a salt-house of 24 shillings. The Bishop of Chester claims a hide and a half, and a salt-house in this manor.

'In Mildestvich hundred the same Richard (Richard de Vernon) holds Wice (Leftwich). Osmer and Alsi held it for two manors, and were free men.

'In Warmundestron Hundret. The same William (William Maldebeng) holds Actune (Acton by Nantwich). This manor has its plea in the lord's hall, and in Wich one house free to make salt.

'In Tunundune hundred. The same William (William Fitz Nigel) holds Heletune (Halton); Orme held it. In Wich there is a house waste.

'In Roelan hundred. The same Gilbert (Gilbert de Venables) holds Herford (Hartford). Dodo held it for two manors as a free man. In Wich one salt-house rendering 2 shillings, and half another salt-house waste.

'In Bochelau hundred. The same Gilbert (Gilbert de Venables) holds Wimundesham (Wincham). Dot held it, and was a free man. There is one acre of wood and a hawk's aery, and one house in Wich, and one border. Randle holds it of the Earl Tatune (Tatton); Lewin held it. There is a house in Wich waste.

'Mildestvic (Middlewich) hundred. Hugh and William hold of the Earl Rode. Godric and Raves held it for two manors, and were free men.

'In the same hundred of Mildestvic there was a third Wich called Norwich (Northwich), which was in farm at eight pounds.

In it there were the same laws and customs as in the other Wiches, and the King and Earl divided the receipts in the like manner. All the thanes who held salt-houses in this Wich gave no Friday's boilings of salt the year through. Whoever brought a cart with two or more oxen from another shire gave 4 pence for the toll. A man from the same shire gave for his cart 2 pence within the third night after his return home. If he allowed the third night to pass he was fined 40 shillings. A man from another shire paid 1 penny for a horse-load. But a man from the same shire paid 1 styca within the third night after his return, as aforesaid. A man living in the same hundred, if he carted salt about through the country to sell, gave a penny for every cart for as many times as he loaded it. If he carried salt on a horse to sell he gave one penny at Martinmas. Whoso did not pay it at that time was fined 40 shillings. All the other customs in the Wiches are the same. This manor was waste when Earl Hugh received it. It is now worth 35 shillings.'

The next references, which are to Nantwich, are interesting, inasmuch as the salt trade has now departed from the locality, the last salt being manufactured at Nantwich about the year 1847.

'*Nantwich.*—In King Edward's time there was Wich in Warmundestron hundred, in which there was a well for making salt, and between the King and Earl Edwin there were 8 salt-houses, so divided that of all their issues and rents the King had two parts and the Earl the third. But besides these the Earl had one salt-house adjoining his manor of Acatone (Acton), which was his own. From this salt-house the Earl had sufficient salt for his house throughout the year. But if he sold any from thence the King had twopence and the Earl a third penny for the toll. In the same Wich many men from the country had salt-houses of which this was the custom :—

'From our Lord's Ascension to Martinmas any one having a salt-house might carry home salt for his own use. But if he sold any of it either there or elsewhere in the county of Chester he paid toll to the King and the Earl. Whoever after Martinmas carried away salt from any salt-house except the Earl's under his custom aforesaid, paid toll, whether the salt was his own or purchased. These aforesaid 8 salt-houses of the King and the Earl in every week that salt was boiled, or they were used, on a Friday rendered 16 boilings of salt, of which 15 made a horse-load. But from Martinmas to our Lord's Ascension these boilings were given according to custom as from the salt-houses of the King and the Earl. All these salt-houses, both of the lord and other people, were surrounded on one part by a certain river and on the other part by a river, and on the other part by a ditch. Whosoever committed a forfeiture within these bounds might make amends either by the payment of 2 shillings or by 30 boilings of salt, except in the case of homicide or of a theft for which the thief was adjudged to die. These last, if done here, were dealt with as in the rest of the shire. If out of the prescribed circuit of the salt-houses any person within the county withheld the toll and was convicted thereof, he brought it back and was fined 40 shillings if a free man, or if not free 4 shillings. But if he carried the toll into another shire where it was demanded, the fine was the same. In King Edward's time this Wich, with all pleas in the same hundred, rendered 21 pounds in farm. When Earl Hugh received it, except only one salt-house it was waste. William Maldebeng now holds of the Earl the same Wich with all the customs thereunto belonging, and all the same hundred, which is rated at 40 shillings, of which 30 shillings are put on the land of the same William, and 10 shillings on the land of the Bishop, and the lands of Richard and Gilbert which they have in the same hundred, and the Wich is let to farm at 10 pounds.

'*Middewich.*—In Mildestwich hundred there is another Wich between the King and the Earl. There, however, the salt-houses were not the lord's, but they had the same laws and customs that have been mentioned in the above-mentioned Wich, and the customs were divided between the King and Earl in the same manner. This Wich was let to farm for 8 pounds, and the hundred wherein it was for 40 shillings. The

King had two parts and the Earl the third. When Earl Hugh received it, it was waste. The Earl now holds it, and it is let to farm for 25 shillings and two wainloads of salt. But the hundred is worth 40 shillings. From these two Wiches whoever carried away bought salt in a wain drawn by four oxen or more paid 4d. for the toll; but if by two oxen 2d., if the salt were two horse-loads. A man from another hundred gave 2d. for a horse-load. But a man of the same hundred gave only a halfpenny for a horse-load. Whoever loaded his wain so that the axle broke within a league of either Wich gave 2 shillings to the King's or Earl's officer, if he were overtaken within the league. In like manner he who loaded his horse so as to break its back gave 2 shillings if overtaken within the league, but nothing if overtaken beyond it. Whoever made two horse-loads of salt out of one was fined 40 shillings if the officer overtook him. If he was not found nothing was to be exacted of any other. Men on foot from another hundred buying salt paid 2d. for eight men's loads. Men of the same hundred paid 1d. for the same number of such loads.

'Flintshire (the detached portion between Cheshire, Denbighshire, and Shropshire).—The same Hugh (Hugh Fitz Osborn) holds Claventone (Claverton, Cheshire), Osmer held it and was a free man. To this manor belong eight burgesses in the city, and they render 9 shillings and 4 pence, and there is a salt-house in Norwich (Northwich) worth 12 pence.'

In the year 1245, when Henry III. was at war with the Welsh, he caused the brine springs of Cheshire to be destroyed in order to cut off the supply of salt to the enemy. In the time of Henry VI. there were 216 salt-houses in operation at Nantwich. In the reign of Queen Elizabeth there were over 200 salt-works of six pans each in Cheshire. In 1671 there were two salt-works in operation at Winsford.

Up to this date the whole of the salt made in Cheshire was obtained by evaporation from the brine of springs; but in 1670 Mr. Adam Martindale communicated to the 'Philosophical Transactions' the fact that in that year John Jackson, in searching for coal on behalf of the lord of the soil, William

Marbury, of Marbury, Esquire, brought up natural salt by an instrument. This first boring was at Marbury Lane, Marston, near Northwich, and it was followed by the sinking of a shaft, and the deposit was worked until the year 1720, when the shaft fell in. Other shafts were sunk, and the working of the rock salt underground at Marston has been continuous until now. Up to the year 1779 rock salt was only obtained from these mines at Marston; but in that year explorations were made at Lawton, which led not only to the finding of the same, but also a lower bed of salt. This discovery led the owners of the Marston mine, in the year 1781, to sink a shaft from their workings by means of a horse-gin, the result of which was the discovery of the lower bed of salt. Explorations have since been made to a depth of 60 yards or so in the strata below the bottom bed of salt, which show thin beds and irregular spheroidal masses of rock salt in the marls. These have not yet been worked.

In 1808 brine springs were chiefly confined to the valleys of the Weaver and the Wheeloch, with a spring at Dirtwich on the borders of the detached part of Flintshire, and a weak spring at Dunham, near the river Bollin.

The most distant point up the course of the Weaver was near Andlem. Here a boring was put down through the slight covering of lias which went into weak brine. Nearer Nantwich there were then several springs in use, but these have been discontinued, partly because the brine was weaker, and partly because the facilities for carriage were not so great as at other places. The most abundant springs were at Winsford, and then at Nantwich and Northwich. Between these two places the inflow of fresh water spoils the successful working of the brine. Springs of weaker brine have been found down the course of the Weaver to Frodsham.

From remote times salt has been the subject of taxation in England. There is a reference to it, we are told, 64 years B.C. The tribute was continued by the Romans when they worked the Droitwich salt springs, and they partly paid the wages of their soldiers in salt. Hence, it is said the origin of the custom at the Eton Montem of asking for salt. Writing in the last century, Dr. Johnson defined this impost to be a hateful tax on commodities not adjudged by the common judges of property. At the beginning of the present century it was computed that 800 tons of salt were wasted annually at the Ashby de la Zouch springs in Leicestershire, because of the tax upon salt, which precluded the possibility of evaporating the brine with the small coal, which was thus also wasted. About the same period the amount of taxes paid by the salt-works of Cheshire and Worcestershire ranged from 1,800*l.* to 20,000*l.* a year. Happily this tax has long been abolished. In the year 1881 the production of rock salt in Cheshire was 166,740 tons, in Ireland 30,891 tons. In the same year the production of white salt from brine was, in Cheshire 1,600,000 tons, Worcestershire 226,000 tons, and Staffordshire 4,000 tons, the both kinds of the aggregate value of 1,149,110*l.*

Referring now to the section, fig. 7, there are in the division 4 two series of beds below the red sandstones that lie at the base of the section. The lowest of these is known as the Lower Mottled group, and it has a thickness, in Lancashire, Cheshire, and Shropshire, ranging from 200 to 500 feet. The next group in ascending order is that of the Pebble beds, which range from 500 to 750 feet in thickness. Above these beds come those shown in the base of the section (4), which have a thickness of about 500 feet.

Above these beds there occurs in Germany a small series of beds known as 'Muschelkalk,' but so far this member of the Trias, as the whole series is termed, has not been recognised in England. The beds No. 2 and 3 of the section form together the lower part of the Keuper, or highest division of the Trias. They consist first in ascending order of a series of flagstones and sandstones of pale buff and red colour, on the bedding surfaces of some of which are beautiful impressions of the footprints of birds and reptiles, good examples of which have been obtained from the Grinshill quarries of the Hawkstone hills. These are succeeded by a similar succession of flagstones and marls, the two series having a combined thickness of from 450 to 500 feet. Fig. 8 gives a general view of this division of the strata as the beds crop out in the ridges of the Peckforton and Delamere hills.

FIG. 8.—GENERAL VIEW OF THE PECKFORTON HILLS, CHESHIRE.

The red marls (1), which form the highest beds of the Trias, attain in Lancashire the great thickness of 3,000 feet. Fortunately for the winning of the rock salt, they do not reach anything approaching this thickness in the Cheshire salt-field, as the following detailed section of the strata overlying the salt deposits near Northwich will show.

DETAILED SECTION OF STRATA SUNK THROUGH AT WITTON, NEAR NORTHWICH, TO THE LOWER BED OF SALT.

	Ft.	In.
1. Calcareous marl	15	0
2. Indurated red clay	4	6
3. ,, blue clay and marl	7	0
4. Argillaceous marl	1	0
5. Indurated blue clay	1	0
6. Red clay, with sulphate of lime in irregular branches .	4	0
7. Indurated red clay with grains of sulphate of lime interspersed	4	0
8. Indurated brown clay with sulphate of lime crystallised in irregular masses and in large proportion . .	12	0
9. Indurated blue clay with laminæ of sulphate of lime .	4	6
10. Argillaceous marl	4	0
11. Indurated brown clay laminated with sulphate of lime .	3	0
12. Indurated blue clay, ditto	3	0
Carried forward . .	63	0

	Ft.	In.
Brought forward . .	63	0
13. Indurated red and blue clay	12	0
14. Indurated brown clay with sand and sulphate of lime irregularly interspersed through it. The fresh water, at the rate of 360 gallons per minute, forced its way through this stratum	13	0
15. Argillaceous marl	5	0
16. Indurated blue clay with sand and grains of sulphate of lime	3	9
17. Indurated brown clay as next above	15	0
18. Blue clay as strata next above	1	6
19. Brown clay, ditto, ditto	7	0
20. The top bed of rock salt	75	0
21. Layers of indurated clay with veins of rock salt running through them	31	6
22. Lower bed of rock salt	115	0
	341	9

On the north-west side of Northwich, the upper bed of salt-rock ranges from 84 to 90 feet. It decreases eastward to 80 feet, and to the south-west it loses 12 to 15 feet in the course of a mile. It would thus appear to form a large lenticular mass, it being probable also that in extending over a larger area it may again thicken. Only 12 to 15 feet near the base of these upper beds are considered pure enough for working.

At Marston, the lower bed is 96 feet thick. In the section above given its thickness is 115 feet; at other mines it has been penetrated to a depth of 117 feet without the bottom being reached.

In the upper 70 feet of the lower bed, the salt occurs in irregular strings and masses, associated with gypsum. It is the lower half that contains the most salt, but of this portion only from 15 to 20 feet near the base are worked.

The rock salt varies from yellow to red and reddish brown in colour, its colour varying with the different proportions of iron it contains. It is crystalline in structure, and fine cube crystals of pure salt occur in places.

The following analysis gives its general composition :—

Chloride of sodium	98·32
Sulphate of lime	0·65
Chloride of magnesium	0·02
Chloride of calcium	0·01
Insoluble matter	1·co

The rock salt sometimes appears in irregular columnal shapes, and sometimes in rounded masses which seem to be compressed into each other. Both the beds of rock salt and those of gypsum abound in the remains of minute forms of life —Estheria and Foraminifera.

The upper marls, with their salt deposits, extend south-east to Audlem, where they are covered with a thin coating of Lias. Brine springs are found here, but the brine is rather weak. As we approach the north-west, towards Nantwich and Beeston, the brine becomes stronger. Eastward there are indications of salt to Congleton and Lawton, and northward to Frodsham. The great mass of the salt, however, seems to be about Northwich, Winsford, and Marston. The lakes of Budworth and Pickmere seem to lie in hollows formed in the outcrops of the salt-beds. Still further to the south-east there are brine springs worked near Weston, north of Stafford. The salt-bearing marls are bounded by two divergent ranges of escarpments, which, starting north-east of Shrewsbury, range the one in the Grinshill and Hawkstone hills to the north-east, and the other beginning in a few isolated hills, at last becomes continuous in the Peckforton and Delamere hills, running to the north-west. In the year 1873, the date of Mr. Dickinson's report, there were at work in Cheshire, Staffordshire, and Worcestershire 19 salt mines and 49 brine springs. Formerly the salt mines were worked by a single shaft of very small diameter. At present, although the diameter of the shafts is small, there are now usually two sunk to the salt-bed worked, and a third for pumping is sunk to just below the base of the drift and soft beds, in order to prevent the surface waters running into the mine. As already intimated, the workings are about 15 to 18 feet high; the roof being supported by thick pillars from 7 to 10 yards square, and from 20 to 25 yards apart, according to the nature of the roof and the overlying strata. The salt rock between is removed by blasting. The drills used are about 8 feet long, pointed at each end, and thick in the middle for handling, hammers not being used. The old fashion, too, of using a straw filled with fine powder for a fuze is also practised. The winding is now done by steam-engines in the usual way, and tram-roads are used underground; no sleepers are used, the rails being fastened to the rock floor of the excavation.

About 300 persons are employed at the salt mines of Northwich and Winsford. The day's work below ground is either from seven in the morning till three in the afternoon, or from eight to four, with two half hours out for meals. The mines are clean and dry, free from carbonic acid gas. There is plenty of head room. The workmen look healthy, and sanitary arrangements are well attended to.

Besides the mines worked in the rock salt beds, there are the brine springs and the pumping arrangements. Indeed, we have seen that this is the oldest industry of the two. The composition of the brines of Cheshire and Worcestershire is given in the following analysis:—

Constituents in 100 Parts Brine.	Cheshire.		Worcestershire.	
	Marston.	Wheelock.	Droitwich.	Stoke.
Chloride of sodium	25·222	25·333	22·452	25·492
Chloride of potassium	—	—	—	—
Bromide of sodium	·011	·020	trace	trace
Iodide of sodium	trace	trace	trace	trace
Chloride of magnesium	—	·171	—	—
Sulphate of potash	trace	trace	trace	trace
Sulphate of soda	·146	—	·390	·594
Sulphate of magnesia	—	—	—	—
Sulphate of lime	·391	·418	·387	·261
Carbonate of soda	·036	—	·115	·016
Carbonate of magnesia	·107	·107	·034	·034
Carbonate of manganese	trace	trace	—	—
Carbonate of lime	trace	trace	trace	trace
Phosphate of lime	trace	trace	trace	trace
Phosphate [illegible]ic oxide	trace	trace	trace	trace
Alumina	trace	trace	—	—
Silica	—	—	trace	trace
	25·913	26·049	23·378	26·397

The rain falling upon the surface finds its way to the rock, entering which by cracks and softer strata it percolates through the beds, and taking up into solution the salt they contain, the water flows on until it finds a natural outlet at a lower level than that at which it entered, or into abandoned salt mines, or into wells sunk on purpose, whence it is pumped up and run into pans, which are subjected to heat, the water passing off in vapour and the salt remaining. This constant abstraction of the solid substances of the strata is producing serious effects in the district. The land sinks, boundaries are removed, roads are turned into rivers, and rivers and canals into roads. In the towns, Northwich especially, the houses are nearly all crooked, and are propping one another up in the spirit of true reciprocity. Occasionally a house altogether disappears, to the great jeopardy of its occupants. What the end of all this will be it is difficult to say. It seems useless to build houses on moving and unstable ground, and it is probable that the district immediately concerned will be given up to agriculture and the production of salt, the dwellings being erected around the outer margin of the shifting area. The usual royalty payable to the landowner ranges from *2d.* to *6d.* per ton, with a ground rent for land occupied. For brine the terms are very various. Sometimes the royalty is merged in an annual rent, at the rate of about 50*l.* for a Cheshire acre—equal to 10,240 square yards. The highest price for salt made from it is about 10*s.* per ton.

Worcestershire.—Droitwich, the centre of the Worcestershire salt industry, lies about twelve miles to the north-east of the city of Worcester. Between the two towns, good sections of the waterstones and other beds underlying the upper red marls are exposed, as are also the strata between the salt strata and the waterstones. Droitwich is situated upon the red marls themselves, and through the incessant outflow and pumping of the brine, the town is in the same danger of falling in as is Northwich. About thirty-five years ago the ground cracked along the axis of the hill to the east of the town very suddenly, so that the sheep feeding close by disappeared in a chasm. Latterly, the houses have been built up on strong wooden frames. Salt springs are also worked at Stoke Prior, to the north-east of Droitwich; on the way to Bromsgrove. At Droitwich the brine springs are in the town, which lies in the narrow valley of the river Salwarp. The sides of the valley rise very steeply to a height of 60 to 80 feet.

The prevailing rock near the surface is a clayey calcareous sandstone of a brownish red colour with spots and patches of bluish green. The brine is reached in the pits or wells at a depth of from 100 to 180 feet. The general succession of the strata in descending order seems to be soil and drift 5 to 15 feet; marl, the rock above described, 30 to 40 feet; hard, flaky or slaty gypsum, formerly called locally, talk, 70 to 120 feet. Immediately in passing through the bed or beds of gypsum the brine is reached, and from the tools dropping 22 inches to 2 feet, it is inferred that there is a river or lake of brine of that depth. When struck the brine rushes up the bore-hole with great force, which indicates that it has its origin at a greater height than that it occupies in the region of the wells. This brine stream rests upon a salt bed. At Stoke Prior the pits are deeper than at Droitwich. A shaft was sunk in the year 1829 to a depth of 153 yards, and the rock salt was mined, but only for a short time. The section of the salt beds here was—

	Yds.	Ft.	In.
To top of rock salt through red marls with gypsum and salt.	121	0	0
Red marl, with veins of rock salt.	0	2	6
Rock salt, red, with from 7 to 20 per cent. of marl . .	13	0	0
Flesh-coloured red marl with veins of rock salt . . .	8	0	0
Rock salt with from 7 to 22 per cent. of marl (not passed through)	10	0	0
	152	2	6

The brine flowing and obtained from the pits is for the most part colourless; sometimes it has a palish green hue, like that of sea-water. It contains from 2,000 to 2,290 grains of salt to the pint. It differs from the Cheshire brine in its containing about 2 per cent. of sulphate of soda. It is also freer from carbonate of lime, oxide of iron, and chloride of lime than is the Cheshire salt. In the Government returns the salt obtained from near Weston, in Staffordshire, is combined with that of Worcestershire. The combined production of white salt from the two counties, at Shirleywich, Stoke Prior, and Droitwich, was in the year 1881 235,750 tons. Seventy years ago the annual production was 16,000 tons, of which the principal part was consumed in England. This production paid a tax of about £320,000. The market price of salt at that time was £31 a ton, of which £30 was duty.[1]

Although not of importance commercially, it may be interesting to note that at Ashby wolds, in the Ashby de la Zouch coal-field, or Leicestershire coal-field, a century ago there were salt springs at a depth of 225 yards from the surface, which contained from 5 to 6 per cent. of salt. Higher up in the series a spring less saturated with brine flowed through the fissures of the coal with a hissing noise, supposed to be caused by the emission of hydrogen gas. Some attempts seem to have been made to utilise these brine springs, but they were abandoned, partly owing to the weakness of the brine and partly to the prejudicial operation of the salt-tax.

North of England.—Formerly salt was produced on an extensive scale about the mouth of the Tyne by the evaporation of sea-water. About two hundred years ago 200 pans were employed for this purpose. In addition to sea-water, brine springs were also to some extent used. At Long Benton Colliery, in Northumberland, the proprietors had early in the present century the exclusive privilege of extracting soda from salt springs in the Coal-measures without paying the usual duty. The principal salt works were grouped about and in some cases gave names to, Howdon Pans, Hartley Pans, Jarrow, and North and South Shields. At the latter place reminiscences of the trade remain in the fact that the town is still divided into East Pan and West Pan wards. Other evidences of the extent of the trade remain in the large hills formed of the ash of the salt-pans.

[1] 'Horner Brine Springs at Droitwich.'—*Transactions, Geological Society*, vol. ii., p. 99.

The production of salt from these sources gradually gave place to the use of salt obtained from Cheshire and Ireland. Sea-water is, however, still largely used to dissolve the rock salt derived from those sources. From 90,000 to 100,000 tons of white salt are decomposed annually on the Tyne in 74,000 to 80,000 tons of sulphuric acid, the result being from 100,000 to 110,000 tons of sulphate of soda.

At the present time there is every prospect of the region south of the Tyne becoming a great salt-producing district. In the year 1864 it was announced that a bed of rock salt had been discovered near Middlesborough. This discovery resulted from a shaft and boring begun on July 4, 1859, by Messrs. Bolckow and Vaughan, for the purpose of securing a supply of water for their works. As this discovery will probably be of great importance to the trade of the district, it will not be out of place to place on record here the details of the strata passed through in this shaft and bore-hole. At the present time it is said that arrangements are being made by the firm named to work the salt deposit.

Particulars of strata sunk and bored through at Middlesborough-on-Tees by Messrs. Bolckow and Vaughan. The sinking was commenced on July 4, 1859.

No.	Description of Strata.	Thickness.		
		Fms.	Ft.	In.
1	Made ground	1	5	0
2	Dry slime or river mud	1	2	0
3	Sand containing water	1	4	0
4	Hard clay (dry)	1	4	0
5	Red sand with a little water	0	1	0
6	Loamy sand with a little water	0	3	0
7	Hard clay (dry)	2	3	0
8	Rock mixed with clay and water	1	5	0
9	Rock mixed with clay (dry)	0	1	0
10	Rock mixed with gypsum (dry)	1	0	0
11	Gypsum with water	0	2	0
12	Red sandstone with small veins of gypsum and water	9	1	0
13	Gypsum rock (dry)	1	0	0
14	Brown shale with water	0	1	0
15	Red sandstone	0	4	0
	Carried forward	24	0	0

	No.	Description of Strata.	Thickness. Fms.	Ft.	In.
		Brought forward	24	0	0
	16	Red sandstone, with veins of gypsum and water	2	0	0
	17	Bluish rock	0	3	0
	18	Red sandstone with water	3	1	0
		Bottom of shaft	29	4	0
Boring.	19	Red sandstone	72	5	4
	20	Red and white sandstone	0	1	6
	21	Red sandstone	35	5	7
	22	Ditto, with clay	0	1	0
	23	Red sandstone	8	4	3
	24	Ditto, and clay	1	3	0
	25	Red sandstone	11	0	5
	26	Strong clay	0	2	9
	27, 28, 29	Red sandstone and clay	6	1	11
	30	Red sandstone with a vein of blue rock	8	1	4
	31	Red and blue sandstone	0	1	5
	32	Red sandstone	1	0	0
	33	Ditto, with thin veins of gypsum	0	1	5
	34	Ditto, ditto	6	3	8
	35	Red sandstone, blue clay, and gypsum	0	1	2
	36	Ditto, with veins of gypsum	14	3	3
	37	Gypsum	0	3	2
	38	White stone	0	0	8
	39	Limestone	0	2	8
	40	Blue rock	0	0	2
	41	Blue clay	0	0	2
	42	Hard blue and red rock	0	0	10
	43	White stone	0	2	7
	44	Dark red rock	0	1	2
	45	Dark red rock, rather salt	1	0	7
Beds of Rock Salt.	46	Salt rock, rather dark	2	0	7
	47	Ditto, very dark	0	4	1
	48	Ditto, very light	0	3	6
	49	Ditto, rather dark	4	3	4
	50	Ditto, very light	7	1	6
	51	Ditto, rather light	1	3	0
	52	Limestone	0	1	0
	53	Conglomerate, limestone, and sandstone, with much salt	1	0	4
			218	5	4
			Or 1,306 feet.		

It will of course depend upon the extent of these salt deposits whether they can be profitably worked, but the thickness is so great that the probability is the beds extend over a large area. Then while some of the beds, as in Cheshire, may not be pure enough for profitable working, there will be thicknesses that may be advantageously mined as in Cheshire and at Wieliezka, as described in the next chapter. An analysis of a sample of the rock salt gave the following results :—

Chloride of sodium	96·63 per cent.
Sulphate of lime	3·09
Sulphate of magnesia . . .	0·08
Sulphate of soda	0·10
Silica	0·06
Oxide of iron	trace
Moisture	0·04
	100·00

The sample was probably one of the lighter varieties.

Following the success attending Messrs. Bolckow and Vaughan's exploration, Messrs. Bell Brothers have had a small shaft, 16 inches diameter, put down by the Diamond Rock Boring Company near to the works at Fort Clarence. This shaft or borehole is 1,200 feet in depth, and it has pierced beds of rock salt 80 feet in thickness, which thus proves the continuity of the salt deposits over some extent of area. It is intended to let fresh water down this shaft and to pump it up as brine.

IRELAND.—Rock salt and salt made from brine have been produced near Carrickfergus for near a century past. The production of rock salt in the year 1881 was 31,730 tons, and of white salt, estimated, 64,000 tons. About a century ago a trial pit was sunk near an old hole on the west side of the Eden, near Carrickfergus, which from time immemorial had been known as the salt hole, but the result was not satisfactory and the shaft was filled up. Subsequently it was noticed that brine oozed out through the filling. A trial was then made by Mr. M. R. Dalway, who discovered a little brine in or near a hard bed in the sandstone about eight yards from the surface. Mr. Dalway then sunk a pit to the east, on the dip of the strata. This shaft was taken down about 177 yards. It reached the hard bed just referred to and a little brine was obtained, but not enough for profitable working. Mr. Dalway also sunk another shaft to a depth of about 233 yards, with a boring below for 17 yards, all in marl, without finding brine. This was about a quarter of a mile from the present salt-pits at Duneme. In the year 1851 rock salt was discovered at Duneme to the east of Mr. Dalway's explorations. As at Marbury Lane in Cheshire, so here the discovery was made in a trial shaft on the Marquis of Downshire's estate, being sunk in the hope of finding coal. Following this discovery other shafts were sunk, and the rock salt has been worked ever since.

The following is a general section of the strata at Dunerne :—

	Yds.	Ft.	In.
Soil	0	1	6
Brown drift with boulders of chalk, limestone, flint, &c. .	20	1	6
Brown, and a little blue marl, with gypsum	224	0	0
	245	0	0
Rock salt	5	0	0
Brown marl and marlstone	2	0	0
Rock salt	31	0	0
Brown marl and marlstone	0	1	0
Rock salt	15	0	0
Total depth from surface	298	1	0

This is Mr. Dalway's mine, whose spirited trials were at last rewarded with success, and it is the deepest salt-mine in the British Islands. In this mine about 40 feet thickness of rock salt is worked. Pillars 14 yards by 12 at the bottom and 12 yards by 10 at the top, are left standing at intervals of 25 yards square apart. At the Belfast Mining Company's mine, in the same neighbourhood, 10 yards of rock salt are worked, and the pillars are 16 yards by 12 at the bottom, and 12 yards by 8 at the top, and the same distance apart. It has been found from experience that with pillars of less strength or frequency the workings crush in.

CHAPTER V.

CHLORIDE OF SODIUM—continued.

Salt Deposits of France, of Switzerland, of Spain—The Salt Mines of Cardona, on the Ebro, near Burgos—The Sea Salt Gardens of the South-west Coast—Analyses of Sea Water—Salt Deposits of Italy, of Germany, Mecklenburg-Schwerin, Hanover, Anhalt, Würtemburg, Bavaria—Salt Deposits of Austria, Salzburg, Wieliezka—Description of the latter Mine—The Sea Salt Gardens of Istria and Dalmatia—Salt Mines of Roumania, of Russia, Solikamsk, Tchapatchi, Orenburg—Salt Lake in the Crimea—Salt Deposits of Africa, of Asia, Caspian Sea, Palestine—Chemical Composition of the Water of the Dead Sea—The Salt Deposits of Persia, of India—Bahadur Khél—Table of Strata—Kohat—Salt Range—North America—"Licks" of Michigan—Petit Anse Island, Louisiana—Nevada—Salt Lake of—Dr. Chas. Darwin's Description of a Saline in Patagonia—Inferences and Conclusions.

IN FRANCE, brine springs occur in several places—at Salies, south of Toulouse, at Salenars and Montmorat, in the Jura, and near St. Maurice. At Arbonne, in Savoy, at an elevation of 7,200 feet above the sea, in the region of perpetual snow, are masses of saccharoid gypsum which are imbued with chloride of sodium, and which become light and porous when the salt has been removed by water. In SWITZERLAND, extensive beds of rock salt associated with gypsum occur in the upper marls of the Trias, or at the base of the Lias, near Bex.

SPAIN.—In this country there are extensive salt deposits, and salt is manufactured from sea-water to a considerable extent. Of the former perhaps the most important is that of Cardona, in the province of Catalonia, in the north-east of Spain.

The district consists of an elevated plain or plateau, through which a valley is excavated, along which runs the river Cardona. Along the sides of this valley, and especially near the town of Cardona, are cliffs of red salt, the town of Cardona itself standing on what has been termed a mountain of salt. The soil of the district is about six inches thick, and it is impregnated with salt, on which vineyards flourish luxuriantly. The workings in the rock salt are in the hill on which the town stands, and in the cliffs in the immediate neighbourhood, for about a mile and a half up and down the valley. The place where the industry centres is an oval valley about a mile and a half long by half a mile wide, extending from the castle of Cardona to a promontory of red salt at the other end of the oval. The salt rock in this promontory is 663 feet high and 120 feet wide at its base. The 120 feet represents the thickness of the bed, the beds being probably highly tilted, and the 663 feet a portion of its extent. So far the workings do not appear to have descended below the bottom of the valley. A rivulet finds its way underground throughout the salt rock, and issues out so strongly charged with salt that in time of floods, when it enters the river Cardona, it kills the fish.

The colour of rocks of the district, as well as of the salt, is red, and the strata abound with crevices and chasms in which cluster stalactites of salt like bunches of grapes. As the sun rises over this scene the effect is described as very beautiful, the mountain of Cardona sparkling as with thousands of precious gems.

As elsewhere, accurate examination shows that the salt deposits occur as large irregular masses in marls and sandstones which themselves are more or less impregnated with salt.

An extensive deposit also occurs between Caparosa and the Ebro; a bed of rock salt 5 feet thick interstratified between gypsum and limestone at Valhirra. Another deposit, at Posa, near Burgos, is said to present indications of volcanic origin, being associated with pumice-stone, and occurring in or near the supposed crater of an ancient volcano.

Notwithstanding these extensive natural salt deposits, the roads and other means of communication have been so defective in Spain that the inhabitants of a considerable tract along the south-west coast cultivate what may be called salt gardens. The sea with each tide is let into a series of ponds, which are constructed for the purpose, in a clayey soil, to a depth of from 2 feet to 6 feet. In the first pond a subsidence of mud takes place; the water is then conducted by a long channel to a second pond, where a further subsidence of mud occurs. Then by a series of channels the water is led through a series of shallow pits, like the settling-pits at a lead mine, only not so deep, in which the salt crystallises, and from which the clear water flows off to the sea.

With this reference to the production of salt from sea-water it will be interesting to notice the result of three analyses of the same, which were as follows to 1,000 parts :—

	1.	2.	3.
Chloride of sodium	24·84	26·66	25·00
" magnesium	2·47	5·15	3·50
" potassium	1·35	1·23	—
Sulphate of lime	1·20	1·50	0·10
" magnesia	2·06	—	0·58
" soda	—	4·66	—
Carbonate of lime and magnesia	—	—	0·20
	31·87	39·20	29·33

In addition to the foregoing there were 6·2 per cent. of carbonic acid, with traces of manganese, iron, phosphate of lime, silica, the bromides and iodides of the metals, some organic matter, and ammonia.

In Italy the volcanic mountain of Cologero, near Sciacca, in Sicily, contains a considerable amount of rock salt imbedded in its layers.

Germany.—Very extensive salt deposits occur in the North of Germany, extending from near Hamburg on the west, through the Grand Duchy of Mecklenburg-Schwerin towards Stettin on the east, and extending to Hanover and to Magdeburg and Stassfurth on the south. The mines at the latter place are said to yield a royalty of 400,000 thalers yearly to the Prussian Government, and to the government of Anhalt 250,000 thalers. The most active operations hitherto have been in the southern half of the area described, but important explorations and discoveries have during recent years been made along the northern portion of the area. Near Jessemtz, about 65 miles from Hamburg, a boring put down by order of the Mecklenburg Government proved saline and gypsiferous strata, containing also potash and other salts, to a thickness of nearly 200 feet. Another boring put down at Lübtheen, near Hagenow, to a depth of 1,496 feet, had at the bottom a thickness of salt of 426 feet. Southwards, at Minden, in Prussia, near Hanover, a boring was started in the Lias, and having passed through the Upper Keuper marls reached the Muschelkalk at a depth of 2,515 feet. From this depth, which, as we have seen, lies at or rather below the ordinary base of the salt deposit, 84 cubic feet of liquid, containing 4 per cent. of chloride of sodium, issued from the boring per minute.

In Würtemburg, in the south of Germany, salt deposits commence at Hall, and are continued south-east through Bavaria to Halstadt, Ischel, and Ebensel in Austria. In Würtemburg these deposits have been long worked near Wimpfen.

A large proportion of salt produced in Germany is obtained from brine springs, and from brine wells by means of pumping.

In some of these, as at Wimpfen, the natural flow or yield of the brine is supplemented by artificial means. A hole is let down to the salt-bed, in which a pump is fixed, but space is left around the pump down which fresh water is sent. This becomes impregnated with salt, and is pumped up brine. The brine derived from this source contains an average of 27·00 of chloride of sodium.

Besides chloride of sodium, which in the natural springs ranges from 0·155 to 9·623, there are in the liquids varying small proportions of other minerals, including the chlorides of potassium, magnesium, and calcium, the carbonates of lime, magnesia, soda, and manganese, and the sulphates of potash, lime, soda, and magnesia. The artificial wells which have been sunk to the salt beds are usually stronger in chloride of sodium than are the waters of the natural springs. Many of the latter are too dilute for profitable evaporation by artificial means. At Salzhausen it takes the evaporation of 339 cubic feet of brine to produce 1 cwt. of salt, and at Schönebeck 19,000,000 cubic feet of brine are required to produce the annual yield of 28,000 tons of salt. In the case of weak brines therefore the greater proportion of the liquid is first removed by gradual evaporation in the open air, so as to bring the solution up to a sufficient strength for profitable working by artificial evaporation.

AUSTRIA.—Salt is produced in the Austrian Empire both by mining and by the evaporation of sea-water. Both industries are monopolized by the Government, and the profits are credited to the revenue. The average production of salt in the empire may be taken as rock salt 554,000 centners, refined salt 1,500,000, sea-salt 220,000, and salt for manufacturing purposes 14,000.

Of the mines the principal ones are those of the Salzberg in the south-west, and of Wieliezka, near Cracow, in the north of the empire.

The deposits of the Salzberg seem to be a continuation of or closely related to those of Würtemburg, in Germany. One of the most ancient and extensive of the mines is situated about 8 miles north of the town of Hall, in the Austrian Tyrol, and on the left bank of the river Inn, below Innsbrück. It lies on red sandstone, marls, and rocks, and at an elevation of 6,300 feet above the sea. There are records extant which show that there were salt-works in operation here early in the eighth century. In all probability these depended upon a salt spring which rises at the foot of the mountain. In the year 1275, so the story goes, Niklas von Rohrbach, *der frommer Ritter*, or pious knight, frequently noticed on his hunting expeditions that the cattle loved to lick certain cliffs of the valley. This led him to taste the flavour of the rock for himself, and finding it rich in salt, he followed up the track until he came to the Salzberg itself, where he inferred from his observations there was an immense supply of salt. Since that time the salt has been worked, first by hewing and later by blasting. Vast chambers, some of which are an acre in extent, have been excavated in the rock, which have been subsequently closed up and filled with water. At the end of a year or so the water becomes impregnated to about one-third its weight with salt, and it is led off by means of conduits to Hall, where it is evaporated. The process of excavation is repeated when necessary.

These salt-beds are continued on to Styria on the north-east, where they are also to some extent worked.

The other great salt mine of Austria is the famous one of Wieliezka, near Cracow, in Galicia. This mine has been worked since the year 1251, and it has still vast reserves of the mineral. Fig. 9 shows a general section of the strata of this mine, and the similarity of the rocks to those in which the great salt deposits of England lie will be at once recognised. The salt occurs in large lenticular and irregular shaped masses, in a red marl rock. These masses are for the most part very pure. In the rock itself there are numerous impressions of vegetable remains, especially in the upper portion overlying the salt deposits. The fig. 9 shows how the deposits are worked, first by a shaft sunk down to the top of the deposits, and then by a series of slanting tunnels which cut through the salt masses. At these points of intersection the latter are excavated, leaving vast chambers, from whose roofs stalactites hang and glisten when the mine is lighted up, and in the sunken floor of the chambers, where the salt mass has been followed down, there are lakes across which the exploring traveller is ferried in boats.

The daily wages of the workmen employed in this mine is about 2*s*., varying somewhat according to the skill and position of the workman. There are also, in addition, grants of firewood and salt. The mine has been worked to a depth of over 400 yards. An American traveller gives the following popular account of a visit he made to this mine some years ago.

'A long winding stair of several hundred steps, neatly covered with boards, led to the first story. Long alleys conducted to the chambers, which, in the course of six centuries,

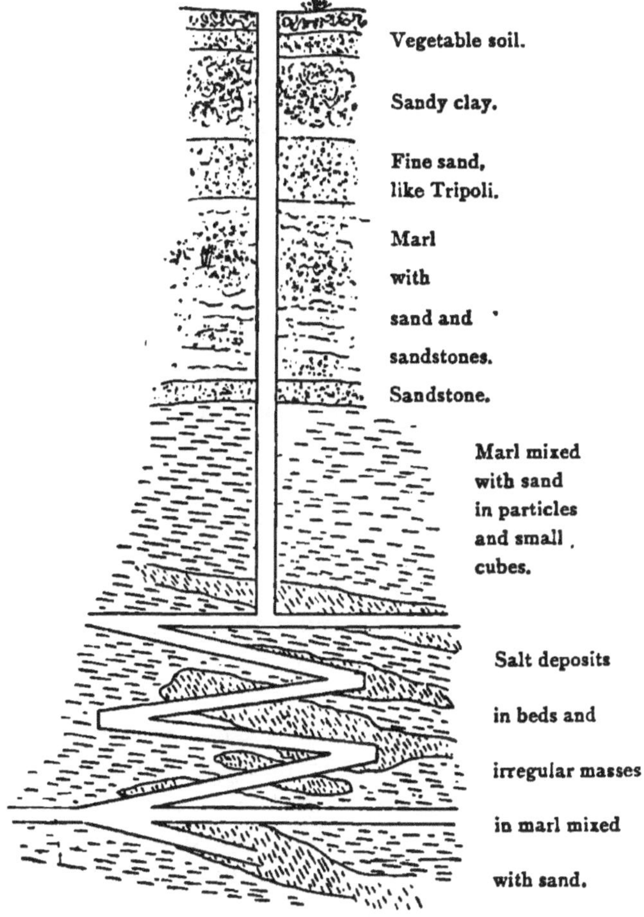

FIG. 9.—SECTION OF STRATA AT THE WIELIEZKA SALT MINES, AUSTRIAN POLAND. Scale 1" = 150 Feet.

have been excavated in the solid salt. These chambers are well proportioned, and present an appearance of cleanliness and neatness which at once reconciles the visitor. No humidity, no closeness, no chilling draughts, but a dry, airy, never varying temperature pervades these subterranean caverns.

The halls up on the first floor have been named after the various monarchs of Poland and Austria, and are decorated with their statues or the monuments erected to their memory. Another chamber is called the chapel of St. Cunegunda, and on the day of her festival high mass is celebrated in the presence of the miners. The largest of the chambers was the concert hall or the theatre. There were the orchestra, saloons, galleries, and from the arched roof above hung a chandelier of salt. Some of the guides ascended to the uppermost tier, and waving their blazing brooms illumined the gloom above and around them. The light falling upon the crystal walls, and the grim shadows trembling and struggling upon the brink of the darkness which reached far beyond into the deep gulf, was marvellously beautiful. Again descending we reached the second story, and threading the long passages arrived at the borders of a lake where a boat and torchbearers awaited us.

'We landed upon the opposite side of the lake, and descended to a chamber immediately beneath; but we were already 600 feet below the surface and thought this quite sufficient. The whole mass above was supported by arches and pillars of salt, as solid and hard as adamant. Some of the latter have been cut away and immense beams of wood substituted in their place. There are no clefts or gaps in the length or breadth of this spacious vault. All is solid and secure, and the idea of accident or damage never occurs to the observer. The rock in its general appearance, and in a doubtful light, resembles our gray granite, except that it has more brilliancy—that kind of brilliancy imparted to the texture of ordinary quarries containing crystallised quartz. Where the water has filtered, crystallisations appear in the forms of cubes and prisms, and where these are seen with the aid of a number of torches, the effect is very beautiful.'

In Istria and Dalmatia, along the east coast of the gulf of Venice and the Adriatic Sea, are important salt works at which the salt is evaporated from sea-water in much the same manner as that practised on the south-west coast of Spain. In this industry 4,400 persons are employed, including 1,700 women and 1,450 children, and the production of salt from this source averages, as we have seen, 220,000 tons a year.

Passing eastwards, into the newly formed kingdom of ROUMANIA, we find five salt mines at work, two of which are penal and are worked by convicts. One of these, Telega, half-way between Ploresti and Sinaia, is descended by a series of staircases to a depth of about 110 feet, the depth it has now reached.

RUSSIA.—This empire produces about 400,000 tons of salt annually. The Government levy a tax of *2s. 8d.* per cwt. upon the article, which amounts to about 12,000,000 roubles a year. This tax is about to be reduced, if the reduction has not already taken place, to *1s. 8d.* per cwt. The principal salt works of Russia are at Solikamsk, on the east side of the Ural Mountains, just on the borders of Russia in Europe, and on the same parallel of latitude as St. Petersburg. Solikamsk is in the kingdom of Perm, which has given its name to our Permian strata; these strata, with the overlying New Red Sandstone and marls, being largely developed there. The production of refined salt at Solikamsk is about 70,000,000 lbs. annually, valued at one halfpenny per pound. The brine is pumped by steam-engines from a depth of from 100 to 150 feet. It is boiled for six hours and left to settle for another fourteen, after which the salt is removed in wooden trays, on which it is left to dry.

Of the 148 salt-works in Russia deriving salt in a similar manner from brine springs, about half are in the Government of Perm. Some of these now in active work have existed since the fifteenth century.

An inferior salt, worth only *1s. 8d.* to *2s.* per cwt., is obtained in a similar manner in the Government of Archangel.

In the Yenotayef district of the Government of Astrachan is the hill Tchapatchi, which is described as being a mountain of rock salt, and from this source a large supply of superior salt is obtained. There are other similar hills more or less made up of salt rock masses in the same district, as well as numerous lakes in which salt is precipitated. Rock salt beds also crop out and are worked near Fletskaya, Fattchita on the borders of the Kirghese Steppes east of Orenburg; and in Siberia there are four imperial salt works, which are worked by convicts. On the northern side of the range of the Caucasus there are salt springs; indeed these abound in south-eastern Russia. The water of a lake near Sympheropol, in the Crimea, is found to contain chloride of sodium 16·12, sulphate of soda 2·444, chloride of magnesium 7·55, chloride of calcium 0·276, and sulphate of potash 0·7453.

Africa.—Before passing into Asia we may just notice that salt occurs largely throughout the continent of Africa. In Tripoli there are extensive lagoons near Bengazi which yield about 200,000 tons of fine salt yearly. The salt rock of Tegara, and those of Had Delfa, in Tunis, are also worked. The mineral occurs in the mountains west of Cairo, and bounding the north of Libya it extends to a great distance. It is also found in solid masses south of Abyssinia. Salt lakes occur to the east of the Cape of Good Hope, which contain at their bottom thick beds of rock salt variously coloured with extraneous matters. Along the coast in the same neighbourhood salt is obtained from sea-water in the way already described.

Asia.—Reference has already been made to the salt mines of Siberia. In portions of the same country on the coast sea-water is subjected to the action of frost, which separates the clear water into ice, leaving a residuum of strongly saline liquid. Salt lakes occur on the borders of the Caspian Sea, which are interesting from their resemblance to others in South America, and as throwing light upon some conditions under which our great Triassic salt deposits were formed. These lakes occupy shallow depressions in the land. The mud on their borders is everywhere black and fœtid. Beneath the crust of sea-salt sulphate of magnesia occurs imperfectly crystallised; the muddy sand is mixed with small strings and masses of gypsum. These lakes are also inhabited by small crustaceans.

In Palestine red rocks containing saline, gypsiferous, and

bituminous matter underlie the Jurassic limestones and strata that cover a large portion of the country and continue southwards into Arabia. Hence it is that the streams running into the river Jordan from the Sea of Tiberias southwards come charged with these various substances, and all flowing into the Dead Sea, which is also bounded by rocks of a similar nature, contribute to the peculiar chemical composition of the waters of that lake. The water of the Dead Sea contains 6·578 of chloride of sodium and 10·543 of chloride of magnesium.

Passing to the south-east we find a considerable salt mining industry carried on in the hills bounding the *Persian* shore of the Persian Gulf, near its entrance into the Indian Ocean. The salt rocks crop out on the sides of the hills. They occur in layers about 4 feet thick, with intervening strata of earthy matter. The general appearance of the rocks is of a reddish colour, and they vary from marl to hard sandstone. There is a good deal of ochre associated with the salt deposits, generally lying above them; so much so that its presence is taken as an indication that salt will be found below.

The chief places of the industry are Kishm Island, Hormuz, Larak, Pohal, Jabel, Bostana, and Hameran. The salt usually occurs in reddish hard granular masses, occasionally in pure white masses. In secondary forms it occurs in stalactitic and saccharoid masses, and in translucent and transparent masses of a cubical form. The red salt is used by the natives for salting fish in connection with the extensive fishery that is carried on along the coast. The finer qualities are sent in native boats to Muscat, whence the salt is exported to Mauritius, Zanzibar, Batavia, and Bengal. The salt rock is quarried by means of powder, and it is afterwards broken by means of wooden and iron mallets. The result of the excavations is the formation of large caverns. At the western end of the district is a beautiful natural cavern formed by the dissolving of the salt out of the rock by the passage of a stream through an original crack or crevice. In the salt springs, which are numerous, sulphurous gases abound, and also crystals of pure sulphur. In some warm springs near the village of Salakh, the water, besides being charged with salt, yields naphtha of a reddish colour, which is highly combustible and burns with a thick smoke, the natives using it for light. The cost of mining the salt is given at 4*s.* 2*d.* per 3,600 lbs., and the price of it at the sea-side at 18*s.* to 22*s.* for the same amount. The whole group of the strata enclosing the salt are considered to be of middle if not more recent Tertiary age.

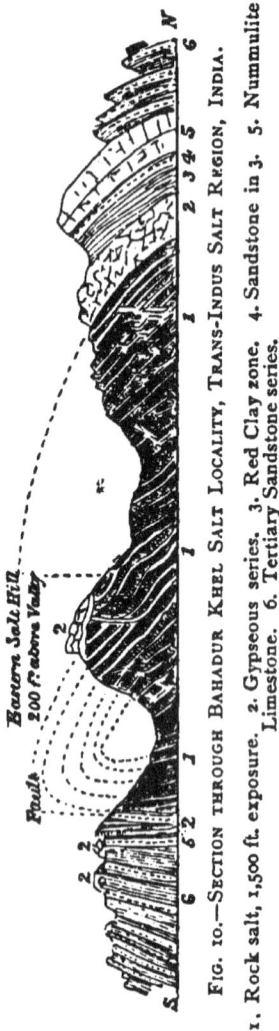

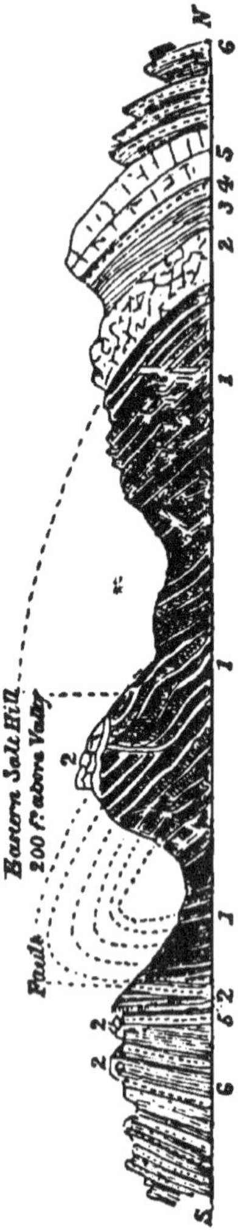

FIG. 10.—SECTION THROUGH BAHADUR KHEL SALT LOCALITY, TRANS-INDUS SALT REGION, INDIA. 1. Rock salt, 1,500 ft. exposure. 2. Gypseous series. 3. Red Clay zone. 4. Sandstone in 3. 5. Nummulite Limestone. 6. Tertiary Sandstone series.

INDIA.—On the west-north-west side of the Punjab, on both sides of the river Indus south of Peshawar, are very important and interesting salt deposits, the more so since they are so near the salt range on the east side of the Indus, which appears to be of Silurian if not an older age, and that on the west side of Triassic or Tertiary age. Of the two, that west of the Indus is of the greatest commercial importance.[1] It comprises an area of 1,000 square miles of country, stretching from outside the British boundary in Afghanistan to near the river Indus, and lying between Bannu and Kohat, but nearer the former place. There are four long and sharp and narrow ranges of hills stretching nearly due east and west; these are steep and rocky. The whole country presents a wild, barren

[1] See *Memoirs of the Geological Survey of India*, vol. xi. 'The Trans-Indus Salt Region,' by A. B. Wynne. With an Appendix on the 'Kohat Mines or Quarries,' by H. Warth.

appearance, with greenish-coloured soil and rocks, varied by bright purple colours and blood-red zones of clay, the white of the adjacent gypsum bands broken by pale orange-coloured or yellowish débris.

The general descending order of the rocks of this region is given by Mr. Wynne as follows :—

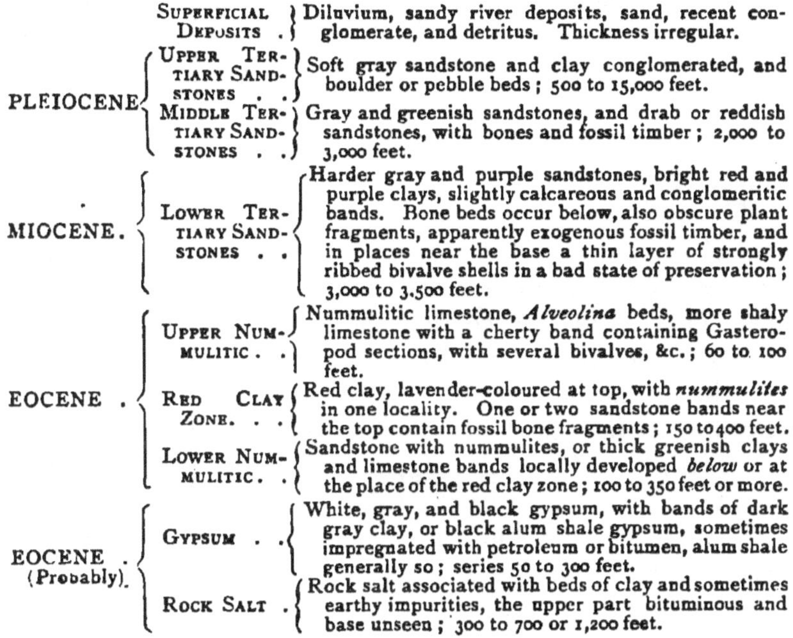

	Superficial Deposits .	Diluvium, sandy river deposits, sand, recent conglomerate, and detritus. Thickness irregular.
PLEIOCENE	Upper Tertiary Sandstones . .	Soft gray sandstone and clay conglomerated, and boulder or pebble beds ; 500 to 15,000 feet.
	Middle Tertiary Sandstones . .	Gray and greenish sandstones, and drab or reddish sandstones, with bones and fossil timber ; 2,000 to 3,000 feet.
MIOCENE.	Lower Tertiary Sandstones . .	Harder gray and purple sandstones, bright red and purple clays, slightly calcareous and conglomeritic bands. Bone beds occur below, also obscure plant fragments, apparently exogenous fossil timber, and in places near the base a thin layer of strongly ribbed bivalve shells in a bad state of preservation ; 3,000 to 3,500 feet.
EOCENE .	Upper Nummulitic . .	Nummulitic limestone, *Alveolina* beds, more shaly limestone with a cherty band containing Gasteropod sections, with several bivalves, &c. ; 60 to 100 feet.
	Red Clay Zone. . .	Red clay, lavender-coloured at top, with *nummulites* in one locality. One or two sandstone bands near the top contain fossil bone fragments ; 150 to 400 feet.
	Lower Nummulitic. .	Sandstone with nummulites, or thick greenish clays and limestone bands locally developed *below* or at the place of the red clay zone ; 100 to 350 feet or more.
EOCENE . (Probably).	Gypsum . .	White, gray, and black gypsum, with bands of dark gray clay, or black alum shale gypsum, sometimes impregnated with petroleum or bitumen, alum shale generally so ; series 50 to 300 feet.
	Rock Salt .	Rock salt associated with beds of clay and sometimes earthy impurities, the upper part bituminous and base unseen ; 300 to 700 or 1,200 feet.

Although the rock salt is in this section grouped with Eocene strata, some diversity of opinion seems to exist as to its exact age. Supposing the intervening strata to be absent, it might still be of Triassic age. Still, although the strata of the district are much contorted, there does not appear to be any unconformity between the undoubted Eocene rocks above and the gypsiferous and saliferous strata below. Fig. 10 illustrates the succession and contortion of the strata.

Owing to this contorted and dislocated condition of the strata, the salt rocks are thrown up to the surface, in some places forming hills and cliffs 200 feet high. The most notable example of this is in the vicinity of Bahadur Khél, where it forms, for the length of a quarter of a mile, high detached cliffs on each side of a small stream valley, forming perhaps the largest exposure of rock salt in the world. Fig. 11, adapted from a beautiful lithograph accompanying the memoir referred to, will give an idea of the appearance of these hills of salt.

In this neighbourhood the salt is very pure. It is of a whitish or gray colour, its texture varying from a highly crystalline mass, the most prevalent form, to a somewhat earthy salt intermingled with blue or grayish finely divided clay. Rarely, minute fragments of gypsum project from the weathered surface of the rock.

The earthy impurities are most common in the western part of the district, where the largest exposures of salt occur, but even here only a few subordinate bands are unfit for working. There is an absence from the salt of potassa and other salts which elsewhere are frequently associated with the salt deposits. The deposits differ from those of Persia already described in that the latter overlie the nummulitic formation, and from those of the salt range 60 to 100 miles to the east of the Indus in that it forms one solid mass from top to bottom, with few exceptions, and also in its colour, nothing like the red or pink salt of the Salt Range being observable in the Kohat district. There is also a difference in age, Silurian rock overlying the salt deposits of the Salt Range.

An analysis of clean salt from the mass at Bahadur Khél gave the following results :—

Chlorine	59·52
Sulphuric acid	1·5
Lime	1·06
Sodium	37·47
Insoluble	·45
	100·00

The salt rock is obtained by quarrying with the ordinary tools and blasting-powders. The ordinary method is as follows. The top soil or débris is removed from a small

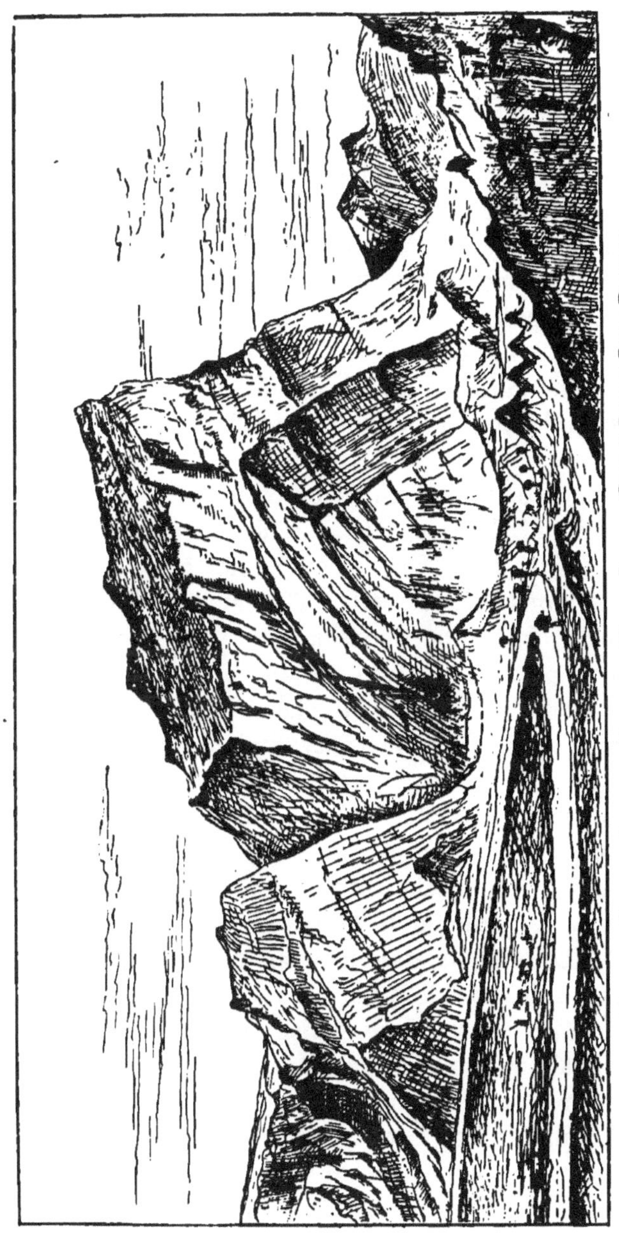

Fig. 11.—Hill of Salt at Bahadur Khel, India. Trans-Indus Salt Region.

central portion of the area intended to be worked, then the salt rock is attacked and removed, the excavation widening downwards as wide and deep as it may safely be taken. Then a fresh ring-like space is cleared around the central opening, the latter being filled with the débris removed, and the rock is again removed; the operation being repeated, the ring-like circumference of quarry extending with each operation, and the central boss or cone of débris increasing correspondingly. The deposits belong to the British Government, who charge a tax or royalty upon the salt removed.

In the Salt Range east of the Indus the salt rock is obtained by ordinary underground mining. Salt deposits in springs of lesser magnitude and importance extend north-eastwards into Thibet.

AMERICA, NORTH.—A chain of mountains extends along the west bank of the river Missouri for a length of 80 miles, by 45 in breadth, and of considerable height. These mountains consist largely of rock salt. The same formation extends into Kentucky, where the deposits are called "licks," because of the licking of the rocks and soil by the herds of wild cattle that once roamed there. In Michigan, in the year 1882, Mr. Crocket McElvoy, of Marine City, sank a well in the neighbourhood to a depth of 1,633 feet, when a deposit of rock salt was entered and penetrated to a depth of 1,633 feet without the tools passing through it. The deposit seems to increase in thickness, but it is reached at an increasing depth as it trends in a south-westerly direction by Inverhuron, Kincardine, and Warwick. The brine is described as pure and strong. An extraordinary superficial deposit of rock salt is described as occurring in *Petit Anse Island*, parish of Calcacren, Louisiana. The island is about two miles in diameter, and the salt deposit on it is known to extend under 165 acres. It is covered with 16 feet of soil. It has been proved to a depth of 80 feet. The salt occurs in solid masses of pure crystals, and it is taken out by blasting. The saltness of Salt River, in Arizona, is due to a considerable stream that, above the junction of the river with the Gila, flows into it from the side of a large mountain.

The bulk of the manufactured salt of North America is obtained from brine springs. Valuable and productive springs are worked in the Syracuse and Salina districts and in Ohio. Some of these arise from a red sandstone whose geological place is said to be below the Coal-measures. There are also the salt lakes of North America, the Great Salt Lake, which has an area of 2,000 square miles and is situated at an elevation of 4,200 feet above the sea. The waters of this lake are described as being a solution of almost pure chloride of sodium.

Rock salt has been recently discovered in Nevada. There are the outcrops of nine beds or ledges, the thicknesses of which range from 30 to 300 feet. The southern termination of these deposits is about 7 miles from the uppermost limit to the navigation of the Colorado River. Some of the specimens are sufficiently pure and transparent as to admit of small print being read through them. In the same State there is an interesting salt lake, the water of which contains about two pounds of salt and soda to every gallon. It is several hundred feet deep. Soda and salt have been obtained from this lake for several years by natural evaporation. The water is pumped into tanks at the beginning of the summer season. It is left in these tanks during the warm summer months, until the frost sets in. When the first frost comes the soda is precipitated in crystals. The water is then drained off into a large pond, where slow evaporation goes on, and a deposit of common salt is obtained. Some beautiful specimens of Gay-Lussite, a compound of the carbonates of lime and soda, and named after the distinguished French chemist Gay-Lussac, are obtained from this lake; the only other locality where they are found being the Lake Maracaibo, in South America. Both in the West Indies and in South America salt deposits and lakes occur. In St. Domingo, about 15 miles from the harbour of Barabona, and between that harbour and the great salt lake of Emiquilla, an important salt deposit occurs. The saline character of the country between the Andes and the Pacific is well known, and important salines occur in Brazil. Perhaps the following description given by the late Dr. Charles Darwin, when he was a young man, and attached to the ship *Beagle* as a naturalist, of the large salt lake or salina 15 miles from the town of El Carmen, in Patagonia, on the south-east coast of South America, latitude 41°, will form a fitting conclusion to the foregoing description of the salt deposits of the world.

'During the winter it consists of a shallow lake of brine, which in summer is converted into a field of snow-white salt. The layer near the margin is from 4 to 5 inches thick, but towards the centre its thickness increases. The lake is 2½ miles long and 1 broad. Others occur in the neighbourhood many times larger and with a floor of salt 2 or 3 feet in thickness, even when under water during winter. A large quantity of salt is annually drawn from the salina, and great piles some hundred tons in weight were lying ready for exportation. It is singular that the salt, although so well crystallised, does not answer so well for preserving meat as sea-salt from the Cape de Verde Islands. The season for working the salina forms the harvest, as on it the prosperity of the place depends. Nearly the whole population encamp on the banks of the river, and the people are employed in drawing out the salt in bullock waggons. The border of the lake is formed of mud, and in this are numerous large crystals of gypsum 3 inches long, while on the surface crystals of magnesia lie about. Worms crawl among the crystals, and flamingoes prey upon them.' The lakes lie in depressions in the grand calcero-argillaceous formation which extends over the Pampas from latitude 20 to 50 south, or in the driftal deposits which lie upon it.

From the foregoing particulars it will be seen that salt deposits occur in strata of all ages, from the Silurian (salt occurring also in older rocks still in a disseminated form) to those now forming. It will also be seen how the conditions under which it has been deposited and also its associated minerals—other combinations of soda, gypsum, magnesia, with oxide of iron as a colouring matter—have been the same through all time. The artificial salt lakes of warm climates and the phenomena of salt lakes in the steppes of south-east Europe and north-west Asia, and in North and South America, indicate to us the ways in which the older salt deposits were accumulated—the thinner seams in shallow lakes, and the thick deposits in deep inland seas and lakes like those of the Caspian and Dead Seas and the deep salt lakes of America. We also see in the description given by Dr. Darwin of the separation of the different minerals in the lakes of Patagonia how the mud settles around the shores, how the disseminated sulphate of lime and sulphate of magnesia gather themselves together in separate masses and crystals, and how in the deeper and stiller portions of the water the chloride of sodium is deposited in a mass with but comparatively little admixture of foreign substances. The same operations go on in arms of the Caspian Sea and in the lakes adjacent. The results of these operations are very similar to the conditions observed in every salt-mine, masses, strings, and crystals of gypsum, other salts of soda and of potash lying usually, as in Germany, upon the salt deposits themselves, and these deposits gathered together in great egg, onion, and thick lense-like masses in the midst of the surrounding strata. Whatever other causes, therefore, may have been at work in the past, according to geological age and local surroundings, yielding at times other minerals besides those commonly associated with the salt masses, we have in phenomena now surrounding us an explication of the conditions under which and the means by which the salt deposits of past times were formed.

CHAPTER VI.

NITRATE OF SODA, BORAX, BARYTA, GYPSUM, FLUOR SPAR, AND ALUM SHALE.

Composition of Nitrate of Soda—Occurrence in German Salt Mines—Deposits of in Peru and Chili—The Desert of Atacama—Statistics of Production—Boron—Boracic Acid—Composition of the Lagoons of Tuscany—Borax—The Tincal Trade of Thibet—Borax in Nepaul, in Iceland, and in Nevada—Barium—Baryta—Sulphate—Carbonate—Sulphate of Baryta in Snailbeach Lead Mine—The Wotherton Baryta Mine, Shropshire—Statistics of Production—Gypsum, its Composition and Varieties—Geological Position—Statistics of Production—Fluor Spar in Derbyshire and in Devonshire—Native Alum—Alum Shale—Alum Industry on the Yorkshire Coast—Description of the Deposit.

NITRATE OF SODA.

NITRATE of soda consists of nitric acid 63·5, and soda 36·5. It crystallises in a rhomboidal form, like carbonate of lime. It burns vividly with a yellow light and deliquesces, in which it differs from saltpetre (nitre).

In most of the German and Austrian salt-mines, nitrate of soda occupies a position overlying the salt deposit, the upper portions of the beds differing in this respect from the lower. The great source, however, from whence it is derived, is the great desert of Atacama, South America, lying between 20° and 27° S. lat., and forming part of the territories of Peru and Chili.

This region is divided into four basins or sub-divisions. The eastern boundary of the whole is the great mountain chain of the Andes, and the basins are divided from each other by mountain ridges running roughly east and west from the Andes to the sea. Starting from the north, the first basin is bounded on the north by the hills Caracoles, Atacama, and Naguayan. This basin communicates with the Pacific by the deep gorge of Negra, near Antojagasta. This is separated from the next region to the south by a range of hills and the peak of Cobre. This second basin is named Cachiyuyal, and towards the sea it opens out into the port of Taltal. It contains the most extensive tracts of level ground to be found in Chili. The third basin is not so large, and it is bounded on the south by the hills that extend to the Cerro Negro and Carrizalillo. It consists more of a series of narrow valleys than of plains to any extent. The fourth basin consists of the dry bed of the Salado River and the undulating stretches of land that bound it. The strike of the underlying strata of the whole region is roughly from north-east to south-west. The oldest rocks, consisting of granites, gneiss, slaty rocks, and limestone, rise up in the mountains to the east. Over these older rocks the newer deposits, consisting of gravel and sand, the waste of the adjacent rocks, are spread, and the characteristic feature of the whole region is its barrenness.

The deposits of nitrate of soda are mainly scattered over the portion of this region lying between 24° and 26° 30′ S. lat. They occur at a little distance from and along the course of ancient river beds. The deposits form layers about 8 to 10 inches in thickness, and they usually underlie a bed of common salt. The richest portions occur near the margin of the deposits. The existence of the mineral under the surface is indicated usually by numerous natural pits leading down to it. The production of nitrate of soda in the Peruvian portion of the region in 1879 was stated at 55,000,000 pounds. It is estimated that the portion of the desert situated in Chilian territory contains enough of nitrate of soda to yield 10,000,000 pounds a year for a century to come. It is probable that, as similar conditions prevail, similar deposits will be found spread over the great salt desert of the Argentine Republic, 25° to 30° S. lat., 64° to 68° W. long.

BORON, BORACIC ACID, BORAX.

Boron is one of the simple elements. It is not abundant in nature, where it is only found in combination with oxygen as boracic acid. It was first discovered by Sir Humphrey Davy in 1807, by exposing boracic acid to the action of a powerful voltaic battery. Gay-Lussac and Thenard afterwards obtained it in greater quantity by heating boracic acid with potassium. In the year 1851 boron was obtained by Deville —1st, in the form of transparent crystals resembling the diamond, but generally of a reddish tint, and in this state he considered it the hardest of all substances known; it scratched the diamond with ease; 2nd, in metallic crystals resembling graphite; and 3rd, as a black amorphous powder.

Boracic Acid consists of three parts of hydrogen, one of boron, and three of oxygen. It is largely obtained from lagoons in Tuscany. The rocks around these lagoons are Cretaceous limestones and Tertiary clays, and are in various stages of decrepitation from the action of the vapours arising from the lagoons. The vapours are often very dense, with a strong sulphuretted hydrogen smell, and the lakes are often restless.

In the year 1818 Mr. François Lardarel, a French gentleman, founded a small establishment for the collection and extraction of boracic acid, and in 1827, being led to practise economy through the great cost of firewood, he turned to account the hot vapours of the lagoons. His trade grew until he had nine establishments within a few miles of Castelnuovo. The production increased from 521 tons in the ten years ending 1828 to 1,831 tons in the year 1859.

Borax, Borate of Soda, is composed of boracic acid 36·58, soda 16·25, and water 47·17. Occurs in white crystals, and has a sweetish alkaline taste. Crude borax, or tincal, is found in Thibet, over extensive districts. Considerable quantities are dug out of the earth, and many of the people, shepherds chiefly, are employed in collecting it. Large quantities are obtained from the lake Pelto, and from another lake distant about one hundred miles to the east of this. This more distant lake is described as being difficult of access, being surrounded by precipitous rocks. It is supplied chiefly by springs, the waters of which hold borax in solution. The water evaporates and leaves a crust of borax covering the bottom of the lake. This process goes on continually, so that there is a constant supply. The ground over large areas is also strongly impregnated with it.

The substance is collected from the sides of the lakes in the months of September, October, and November, when it is carried by flocks of sheep, about 30 pounds on a sheep, to the villages, where it is packed in bags woven by the shepherds, and further carried at a very slow rate over difficult roads to Moradabad, where it is bought by native merchants and sent by them to Calcutta.

A deposit has also been discovered in Nepaul, consisting of very fine crystals and comparatively pure. A lagoon has also been discovered in Iceland, and, as I write, the prospectus of a company just formed to work it is issued.

A similar lake exists in Esmeralda country, Nevada, United States of America, in a valley known as Teel's Marsh. This was discovered as lately as 1873, and a great industry has arisen in connection with it. It appears as a soft clayey deposit, and it is said, after removal, to renew itself in two or three years.

BARIUM, BARYTA.

Barium, another of the simple elements and the metallic base of barytes, was also discovered by Sir H. Davy, in the year 1808. It is a white metal like silver, and is fusible below a red heat. It takes its name from *βαρύς*, heavy, on account of the density of some of its compounds.

Baryta consists of about equal parts of barium and oxygen. It is a grey powder with a specific gravity of about 4.

Sulphate of Baryta (*Heavy Spar*).—Composition : baryta 66, sulphuric acid 34. H. = 2·5 to 3·5 ; specific gravity 4·3 to 4·8. Of a white colour, frequently tinged yellow, red, or brown; occurs in compact, granular, fibrous, and columnar masses; is translucent to transparent, with a vitreous lustre. It decrepitates before the blow-pipe, and fuses with difficulty.

CARBONATE OF BARYTA (*Witherite*).—Composition: baryta 76·6, carbonic acid 22·4. H. = 3 to 3·75; specific gravity 4·29 to 4·35. Crystallises in six-sided prisms with terminal pyramids, and exists also in fibrous or granular masses. Colour, white, yellowish, and grey, and a rather resinous lustre. The crystals are usually white and transparent.

Baryta occurs in the Llandeilo and Bala beds of the Lower or Cambro-Silurian strata. It is found associated with lead ores. At the Snailbeach lead-mine, in Shropshire, it occurs in beautiful crystalline forms. It is, however, in the higher beds of these strata, where the lead fails, that baryta is most abundant. The Wotherton mine, on the borders of Shropshire and Montgomeryshire, is one of the most productive baryta mines in Great Britain. Here the lode is from 3 to 30 feet wide. It crosses a band of hard rock of considerable thickness diagonally from east to west, and it is in this hard rock that the lode is most productive. When it passes at each end into soft slate or shale, it becomes disorganised, and the baryta is in a more scattered form. In the productive portion the lode consists of great masses of pure white sulphate of baryta, mixed with others coloured yellow. Some very perfect crystals occur in cavities in the masses, some of which are the more interesting because they show the method of their growth by accretion, in the layers coloured by different metals. There are also fine crystals of carbonate of lime which are similarly coloured. There are also here and there spots and nests of the sulphides of copper and lead. This mine yielded 3,328 tons in the year 1881. Three mines in Northumberland and Durham yielded 5,434 tons, including a considerable proportion of carbonate of baryta. One in Yorkshire, Raygil, gave 2,017 tons, and forty-four in Derbyshire a total of 5,140 tons. The baryta from the four last-named counties came from the strata of the Carboniferous Limestone, in which it was also associated with lead-ores, where it is found at a depth of from 70 to 100 feet below the ordinary gypsum. In Great Britain and Ireland there were, in the year 1881, seventy-two mines, with a total production of 21,313 tons. The sulphate of baryta is ground for the manufacture of paint, and the carbonate, which is a poison, is used among other things for the destruction of rats.

GYPSUM—SULPHATE OF LIME.

We have already noticed carbonate of lime as constituting the bulk of limestone rocks, but another form, sulphate of lime, occurs abundantly in nature. The general composition of gypsum is, lime 32·6, sulphuric acid 46·5, water 20·9. H. = 1·5 to 2·0, and specific gravity 2·31 to 2·33. In a pure and crystallised state it is clear and translucent, with a pearly lustre; but according to the degree in which it is mixed with other minerals, it is grey, yellow, brown, and black, and is opaque. It crystallises in right rhombic prisms with bevelled edges. Its varieties are—

ANHYDRITE (*Anhydrous Sulphate of Lime*) differs from ordinary gypsum in that it does not contain water. Chemical composition: lime 41·2, sulphuric acid 58·8. Occurs at the salt-mines at Bex, in Switzerland, where it is found at a depth of from 70 to 100 feet below the ordinary gypsum; Hall, in the Tyrol, at Ischil in Upper Austria, and at the great salt-mine of Wieliezka, in Austrian Poland. It is sometimes used as an ornamental stone.

Fibrous Gypsum or *Satin Spar*, composed of fine white fibres.

Radiated Gypsum, having as its name indicates a radiated structure.

Selenite, which includes the foliated transparent gypsum.

Snowy Gypsum and *Alabaster*, the latter being the name by which the massive form of gypsum is known.

Sulphate of lime occurs in most geological formations, from the oldest Silurian to the newest Tertiary. It is, however, most abundant in the upper or Keuper division of the Triassic strata. We have already seen, in treating of salt, how largely it is associated with that mineral in almost every salt district described, occurring generally in nests, pockets, and irregular masses above the salt deposits, and in one instance, that of Droitwich, forming a continuous bed 40 to 100 feet thick above the liquid brine.

In the year 1881 there were sixteen gypsum mines at work in Great Britain and Ireland. These returned a total production for the year of 79,498 tons, of the estimated value 23,329*l.* Of this quantity, three mines in Derbyshire yielded 12,928 tons, nine mines in Nottinghamshire yielded 49,604 tons, three mines in Staffordshire 7,456 tons, and one mine in Sussex 9,510 tons. Formerly the great source of gypsum was the Chellaston Plaster Mine in Derbyshire; but its production had fallen off in the year 1881 to 790 tons.

It is estimated that from 30,000 to 40,000 tons are used annually in Great Britain, chiefly in the Staffordshire potteries for making plaster moulds; hence it is often called 'potters' stone.'

In France, gypsum is worked chiefly at Montmartre and Pantin, in the Paris basin of Tertiary strata. It is harder than the English gypsum, and as plaster of Paris it is more valued on this account.

As alabaster, gypsum is worked in several parts of Italy. The purest is that of Val di Marmolago, near Castellina, thirty-five miles from Leghorn, and it is much used for ornamental purposes. A fine variety, resembling white wax, is also obtained from Valterra, and a granitic variety is obtained from Carrara.

In America, selenite and snowy gypsum occur in New York, near Lockport; and in the Mammoth Cave, Kentucky, it occurs in beautiful imitations of flower, shrub, and tree foliage.

FLUOR SPAR.

FLUOR SPAR (*Fluoride of Lime*) is composed of: lime 51·3, and fluorine 48·7. It occurs in compact and granular forms; is sometimes fibrous. It ranges in colour from white through yellow, light green, purple, blue, and more rarely rose red. The colours are usually bright, and in the massive varieties they are frequently banded. It frequently forms the gangue or matrix of the metallic ores in mines, especially in lead-mines in limestone, and it is found occurring in veins in the older gneissic, granitic, and slaty rocks.

There are three mines in England producing fluor spar—two in Derbyshire in the Carboniferous limestone, yielding in 1881 122 tons, and one, the Tamar Silver Lead, in Devonshire, worked in the older slaty rocks, which produced in the same year 250 tons, the value being about 14*s.* per ton.

NATIVE ALUM—ALUM SHALE.

Native alum occurs in silky fibrous masses, and also in octahedrons and efflorescent crusts. The usual composition is: 24 parts of water to 1 part of sulphate of alumina, and 1 part of some other sulphate. In the common alum of the shops, potash alum, this sulphate is a sulphate of potash. In soda alum it is sulphate of soda; magnesia alum, sulphate of magnesia; ammonia alum, sulphate of ammonia; iron alum, sulphate of iron; manganese alum, sulphate of manganese. Alum is manufactured largely in England from alum shale, of which in the year 1881 4,950 tons were raised in Lancashire, 52 tons in Warwickshire, 2,651 tons in Yorkshire, and 8 tons in Scotland.

The raising of alum shale, and the production, is an old industry in Yorkshire. In 1460 Sir Thomas Challoner brought over a workman from France to carry out in England at Guisbro' the then secret process, the monopoly of the trade being in the hands of the Pope.

The works afterwards passed into the possession of the Crown, and were declared to be a royal mine. They were subsequently let to Sir Paul Pindar, at a rental of 15,000*l.* a year. He employed 800 persons and made large profits, the price being then 26*l.* per ton. During the Commonwealth the mines were restored to their original owners, and five manufactories were at work.

The chief quarrying of alum shale now takes place near Whitby, in Yorkshire. Overlying the deposit, which occurs in Liassic strata, is a bed of hard compact stone called *Dogger.* Below this comes a thick deposit, between two and three hundred feet thick, of hard shaly clay, of a bluish grey colour. At a considerable depth from the surface this is as hard as ordinary slate, but on exposure to the weather the hardest portions crumble and decompose. The upper part of the deposit is soft and has an unctuous feel. Lower down the deposit becomes dull and earthy, with an admixture of sand and of ironstones, but below it again assumes its softness and unctuousness. The parts of the deposit which are more earthy than slaty yield the most alum. The whole deposit abounds with sulphur in the form of iron pyrites, but sulphur is more abundant in the upper part, which is the most valuable part of the deposit.

CHAPTER VII.

PHOSPHATE OF LIME.

Phosphorus—Importance in Vegetable and Animal Life—Use in Agriculture—Mode of Occurrence in Nature—Phosphoric Acid—Apatite—Other Forms of Phosphate of Lime—Professor Henslow and the Coprolites of Suffolk—Modes in which Phosphate of Lime occurs in Nature—The Apatite Deposits of the Laurentian Rocks of Canada—Their Range and Manner of Occurrence—Particular Examples—Analyses—Particulars of Mining—The Apatite or Phosphate Deposits of Norway—Range—Geological Age—Various Modes of Occurrence—Particular Examples—Rutile—Rock Dykes—Analyses—Difficulties of Dressing—Mining Particulars.

PHOSPHORUS is one of the simple elements. It was discovered as such in the year 1669, by Brandt, of Hamburg, and in 1769 Scheele discovered its presence in the bones of men and animals.

Since the last date it has been found to be a most important and essential ingredient of the brain, and to be necessary to the nervous system generally. It is therefore at the present time largely employed in the preparation of medicines for ailments affecting these parts and functions of the human body.

It is present also in considerable proportions in plants, and books on agricultural chemistry usually contain numerous details of the quantities found in the various plants and roots that are used for human food.

When it is reflected how much phosphorus must be extracted from the soil every year to make the bones and tissues of all the living things that grow out of and feed upon the earth, it will be seen how necessary it is that at least as much phosphorus as is extracted from the soil should be returned to it from time to time if we would avoid its utter exhaustion. In the attempt to do this we have suggested to us the whole question of the preparation of chemical manures, into the composition of which this substance largely enters.

Phosphorus is not found in a pure form or free in nature, but always in combination with other substances, chiefly lime and oxygen. In the mineral state in which we have now to consider it, it is found in combination with lime in the following forms, in which it is known as phosphate of lime.

In all these forms the strength and value of the mineral is calculated according to the amount of phosphoric acid there is in combination with the other minerals. The composition of phosphoric acid is as follows: phosphorus 31 parts, oxygen 64 parts, hydrogen 3 parts.

Apatite.—Chemical composition: phosphate of lime 91 to 92, chloride of calcium 0·0 to 0·42, and fluoride of calcium 4·6 to 7·7. This mineral has a specific gravity of 3·16 to 3·25, that is to say it is about three and a quarter times as heavy as water, and its hardness is = 5. In appearance it is often transparent; its natural colour is a creamy white, but it is generally tinged yellow, light green, grey, and blue, by the admixture of various substances that enter into its composition. Its varieties are known as—

Phosphorite, which is the general rocky or massive form in which the mineral is found, as described more particularly further on.

Magnesia Apatite, which contains 7·74 of magnesia. This variety is found at Kusinsk, in the Ural Mountains, but magnesia often enters into the composition of phosphatic deposits.

Moroxite, which is apatite of a greenish blue colour and opaque character, from Arendal, in Norway.

Asparagus Stone, which occurs in translucent crystals of a reddish colour, and is found in the Zillerthal, in the Tyrol.

It will be seen in the following pages that the commercial

value of the phosphates of lime of commerce is principally determined by the amount of phosphoric acid they contain in conjunction with lime.

It was in the year 1842, when on a visit to Felixstow, in Suffolk, that the attention of the late Professor Henslow was directed by a countryman to the existence of curiously shaped nodules or concretions in the red crag of that neighbourhood. On examination the Professor found that these nodules contained a considerable proportion of phosphate of lime, and at the meeting of the British Association in 1845, he suggested the value of these substances in their application to agriculture.

The result has been extensive mining operations in the original deposits, and also in other deposits of the same substance which have subsequently been discovered in other parts of this country and in various countries of the world, together with a corresponding growth in the manufacture of chemical manures. To such an extent has this industry grown within the last thirty-five years, that in the year 1875 there were raised in England and Wales 250,152 tons of phosphatic nodules and substances of the value of 628,000*l.*, besides which we imported 100,258 tons. In 1877 the home production had fallen down to 69,000 tons, our manufacturers depending chiefly upon foreign supplies. The port of Charlestown supplied in that year no less than 170,000 tons, with about 70,000 tons derived from other countries.

Phosphate of lime is contained in rocks of all ages and of almost all textures. It occurs as chemically mixed throughout the mass of rock, as collected in a purely mineral form in cracks, cavities, and layers of the strata, and also as gathered into and forming a good part of the fossil remains contained in strata, as well as in beds in which the phosphatic matter of ancient sea organisms has been re-deposited upon the old sea floor.

I proceed to describe the principal phosphatic deposits of the world, beginning with the oldest stratigraphically, and ascending to the newest deposits, noticing as we proceed the intervening strata in which the mineral is more widely and sparsely diffused.

PHOSPHATIC DEPOSITS OF CANADA.

At the base of the whole series of stratified rocks lie the Laurentian strata of Canada, the equivalents of portions of which are found in the Western Highlands of Scotland, with possibly outcrops of the same strata in the promontories of Carnarvonshire and Pembrokeshire, in Wales.

Fig. 12 will afford an idea of the succession of these lowest

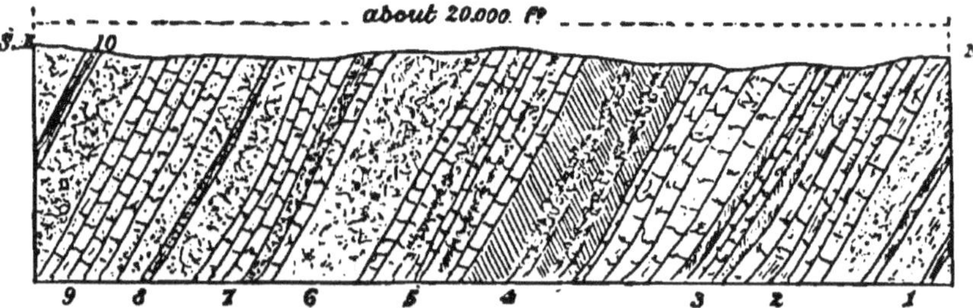

FIG. 12.—ORDER OF LAURENTIAN STRATA OF CANADA WITH PHOSPHATIC DEPOSITS.

1. Speckled hornblendic or pyroxenic gneiss and greenish hornblendic slaty rocks. 2. Crystallised limestone, White Lake and Bolton's Creek band. 3. Bands of white and bluish-grey limestones (Upper Sharbot Lake), interstratified in lower part with beds of mica slate. 4. Black hornblendic slates of quartzose gneiss. 5. Lower Sharbot limestone with interbedded bands of gneissic and hornblendic rock. 6. Granitic gneiss of great thickness (8,000 to 9,000 ft.), with apatite. 7. Crow Lake, Rock Lake, and Silver Lake limestone with deposits of apatite near its base. 8. Gneissic strata of great thickness with a thin interbedded band of limestone. 9. Bols Lake and Tay River limestone. 10. Gneissic strata of great thickness.

known strata of the earth's crust, as they course from north-east to south-west through the provinces of Ottawa and Ontario, in Canada, and the explanation of the figure will give a general view of their mineralogical characteristics.

The whole of this vast volume of strata undulates from north-west to south-east in synclinal troughs and anticlinal ridges, and it is where the ridges of the granitic gneiss, No. 6, with its overlying limestone, No. 7, come to the surface that the phosphatic deposits have been chiefly found and worked. The region where the deposits have hitherto been most proved lies for some distance on both sides of a line drawn from Prince

Edward Peninsula on Lake Ontario, north easterly through the counties of Frontenac, Leeds, and Renfrew, in the province of Ontario, to the river Ottawa, below the island of Callumette, and thence through the county of Buckingham, in the province of Ottawa.

The deposits occur in three forms. First, as beds of irregular thickness, interposed between the almost vertical strata. Secondly, as veins or lodes whose general direction is north-west by south-east or at right-angles to the run or strike of the strata; and thirdly, as superficial deposits in the detritus that covers the upturned edges of the rocks. These are the result of the decomposition of the exposed portions of the strata, the blocks and rough crystals of apatite having fallen out of the disintegrating mass of rock, the fragments and remains of which now form the loose material surrounding them.

The discovery of these phosphatic deposits is of comparatively recent date, and to the close of the year 1874 the workings on them partook more of the nature of preliminary trials than of systematic workings. There were at that date one hundred and forty-two openings made on deposits in North Burgess, the general position of most of which is shewn on the accompanying sketch-map, fig. 13.

These openings consisted chiefly of trenches 10 to 20 feet long by 4 to 10 feet wide, cut through the superficial soil and decomposed part of the solid strata below. Many of these openings revealed phosphate in the loose covering, and a good many tons were shipped. In this position the phosphate consisted of loose masses, embedded in a micaceous pyroxenic débris, surrounded by a good deal of calcareous matter, together with carbonate of baryta. These superficial workings were uncertain, not continuously profitable; they were soon exhausted, and consequently abandoned.

Following, however, the indications they afforded down to the solid rock, many of them revealed both beds and veins of phosphate of lime, some of which have proved continuously and profitably productive.

It was difficult at first to distinguish between a bed and a vein, but as the bearing of the strata became understood, the

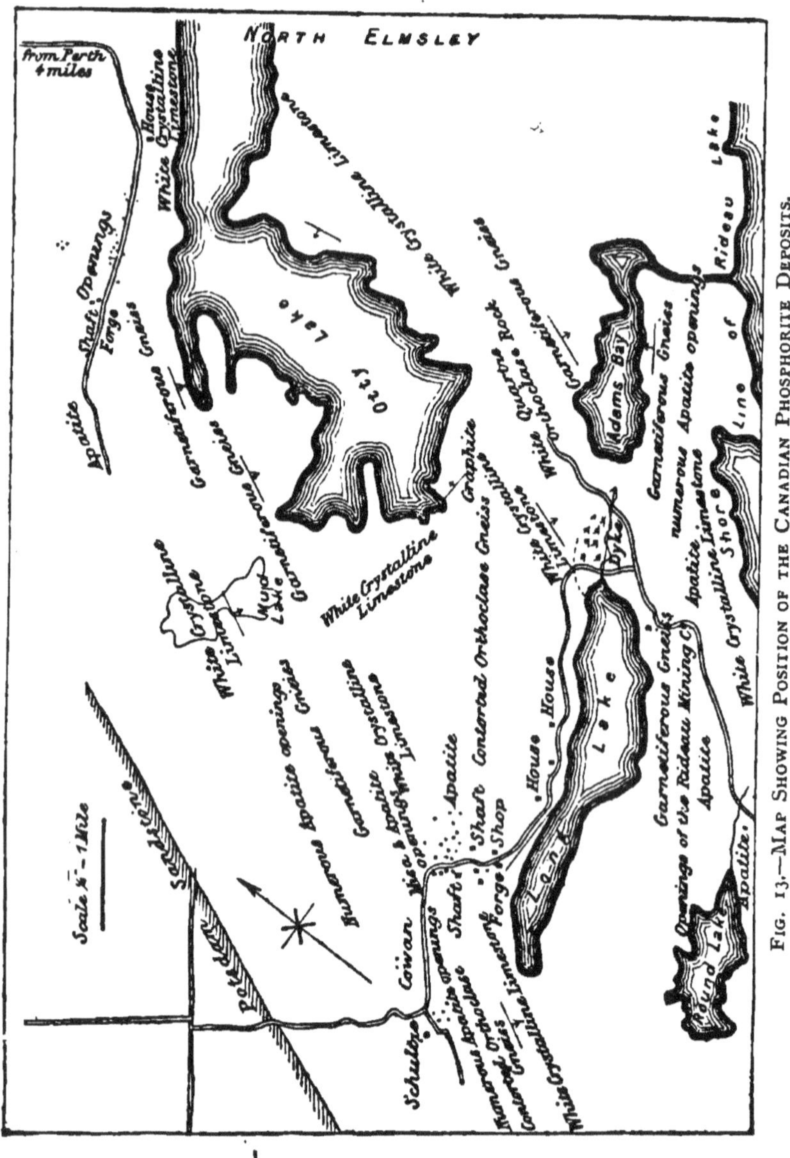

FIG. 13.—MAP SHOWING POSITION OF THE CANADIAN PHOSPHORITE DEPOSITS.

veins were readily distinguished by their contrary direction to that of the strike of the stratification.

In order to afford a view of the character of these more

permanent workings, I will give a few selected descriptions of them.

1. An opening 10 ft. long, 7 ft. wide, and 15 ft. deep, revealing at the bottom a bed of green massive apatite, varying in thickness from 1 to 2 ft., and enclosed by dark quartzose and micaceous hornblendic gneiss. This bed was struck along its course westward in two places within a chain's length, where it showed the same character.

2. A trench 30 ft. long, sunk down to a bed of calcite, with a north-east and south-west course that contained crystals of apatite of large size.

3. A similar trench sunk on a parallel bed of calcite of a red colour, in which were crystals of apatite grouped together in the midst of the carbonate of lime.

4. Two openings on a bed running east and west, and averaging 10 in. in thickness; a beautifully pure bed of apatite of excellent quality.

It may be observed here that the apatite deposits in carbonate of lime have not proved so continuous as those enclosed in the gneissic rocks. To the foregoing I will add some illustrations of the character of the veins.

1. An opening 10 ft. square sunk down to an irregular vein of green apatite running in a north-west direction and enclosed in a rock which is a mixture of carbonate of lime, felspar, mica, and pyroxene.

2. An opening 8 ft. long, 4 ft. wide, and 3 ft. deep, exposing a vein of green phosphate, bearing north-west, which had a thickness varying from 3 in. to 2 ft., and which was enclosed in a rock of quartzose granitic nature.

3. A pit 10 ft. long, 4 ft. wide, and 6 ft. deep, exposing a vein of green apatite 6 to 18 in. wide. The apatite is here associated with good-sized crystals of whitish coloured mica.

4. An opening 35 ft. long, 4 ft. wide, and 6 ft. deep, shewing a vein of green apatite of a thickness of from 2 to 3 ft. in syenitic rock.

5. A shaft 30 ft. deep, in a rock of granitic gneiss, down

to a vein of apatite varying in thickness from 18 in. to 17 ft., along which a level had been driven for a distance of 85 ft. In this distance two smaller veins were seen to branch out of the main one. The phosphate occurs in pockets and bunches in the vein, the enclosing rock being a quartzose gneiss. From these workings about 450 tons of high class phosphate of lime had been obtained. The pockets and bunches of apatite are usually connected with each other by a leader or string of the mineral, but they are often quite cut off and separate from each other.

The examples just given will afford an idea of the nature of the phosphatic deposits in Ontario, and the following description of a deposit in Ottawa will show the similarity of their structure at the extreme point north-east at which the deposits have been worked.

The deposit is on the eighteenth and nineteenth lots of the twelfth concession of Buckingham. It is situated on the Rivière du Lièvre, where the rocks rise in a bold cliff about 100 ft. high. This escarpment of rock is of a similar character to those already described—granitic gneiss. It is intersected by numerous veins of green crystalline apatite, which frequently occur in aggregations or clusters of large-sized crystals. One of these, which is before me as I write, is as thick as a man's thigh. A portion of it yielded 93 per cent. of phosphate of lime. The crystals are cemented together by a readily crumbling matrix of cream-coloured carbonate of lime, in which smaller crystals of apatite are thickly disseminated. The proportion of phosphoric acid is usually larger in these crystals than in the rough granular masses.

Care is required in separating these crystals from the enclosing substances, which in colour and texture so closely resemble the mineral. An instance occurred in which a work was carried on for some considerable time, during which several thousand tons of pyroxene, which closely resembles the phosphate, was stored for shipment. Instances are not uncommon, too, where the phosphate has been thrown upon the waste-heaps. This last mistake has arisen from the variable character and colour of the apatite, which certainly here justifies its name, which signifies to deceive.

When first discovered the colour of the mineral was green, and all substances not green were rejected, but it is now recognised in every shade of colour, the next principal variety being red. It is, as I have said, often crystalline; it is also roughly lamellar, and from this form it passes into that of a granular condition, known as sugar phosphate. It also occurs as a very close-grained compact rock, which has formerly often been thrown away as useless. Several blocks of almost pure phosphate, weighing upwards of a ton each, were shown at the Paris Exhibition.

The following analysis by Dr. Voelcker will show the general composition of the Canadian phosphates, and it is seldom that any of a less proportion than 65 per cent. of phosphate of lime are shipped from Canada.

	No. 1.	No. 2.	No. 3.	No. 4.	No. 5.	No. 6.
Moisture, water of combination, and loss on ignition	·62	·10	·11	1·09	·89	1·83
*Phosphoric acid	33·51	41·54	37·68	30·84	32·53	31·87
Lime	46·14	54·74	51·04	42·72	44·26	43·62
Oxide of iron, alumina, fluorine, &c.	7·83	3·03	6·88	13·32	12·15	9·28
Insoluble silicious matter	11·90	·59	4·29	12·03	10·17	13·50
	100·00	100·00	100·00	100·00	100·00	100·00
* Equal to tribasic phosphate of lime	73·15	90·68	82·25	67·32	71·01	69·35

The distinguishing characteristics of the composition of Canadian phosphates are the absence of carbonate of lime, the scarcity of iron and alumina, and the presence of fluorine. They are also rather hard, and somewhat difficult to grind to the desired degree of fineness.

In the early stages of the industry, when the phosphate has been picked out of the loose stuff near the surface, it has been known to be mined for from *2s.* to *2s. 6d.* per ton, and large quantities have been obtained at prices varying from 5*s.* 6*d.* to 10*s.* per ton. But as the deposits are followed down into the hard gneissic rock the cost of mining is considerably increased. The rock is very hard, and it soon blunts the drills; against this, however, must be set the fact that frequent joints in the rock facilitate operations. The veins are worked by open cuttings, and, where favourable, by means of tunnels, from which the veins are followed and stoped away upwards overhand.

The following is an estimate of the cost of working made by Mr. Alexander Garret, of Ottawa, in connection with the Rivière Lièvre deposit, before referred to, which, although it may be subject to variations caused by time and by special mineral conditions, will afford a favourable idea of the cost of mining Canadian phosphates. Quantity supposed to be raised when the mine is fairly opened out, 5 tons a day.

	$	Cents.
Four men at 120 cents each per day	4	80
One cart, horse, and boy	2	25
Assorting	0	75
Loading on raft	0	50
Freight on raft to Buckingham	2	00
,, to steamboat-landing on the Ottawa	7	50
Wharfage	0	60
Loading on barge for Montreal at $2 50 cents per ton	12	50
Discharge into vessels at 10 cents per ton	0	50
Commissions and insurance on 5 tons	3	00
Powder and fuze	0	20
Interest and contingencies	0	50
Loss in transit	0	75
	35	85

or $7 17 cents per ton.

It is assumed in the calculation that the phosphate is equal to 80 per cent., and that it is worth in Montreal $20 a ton, so that the cost and profit would stand thus:—

	$	Cents.
Value	100	00
Cost	35	85
Profit	64	15

or nearly 2*l.* 14*s.* per ton.

As, however, the value of the phosphate in Liverpool at

1*s*. 4*d*. per unit would be only 5*l*. per ton, and the freight from Montreal to Liverpool would average 25*s*., the value of the mineral in Montreal as given above is placed too high.

An estimate at another mine is given as follows:—

	£	*s*.	*d*.
Cost of mining per ton	1	5	0
Cartage to river	0	4	0
River carriage to Montreal.	0	4	0
Freight from Montreal to Liverpool . .	1	5	0
Cost in Liverpool	2	18	0

The amount shipped from Montreal in 1883 was 17,840 tons.

As regards the cost of mining, this estimate seems to be most correct of the two. The cost will, however, differ much in different localities, owing to the varying thickness of the beds, the frequency or otherwise of the pockets, their size, and other considerations. It will also vary at different times at the same mine. Let us now notice deposits of a similar nature in Norway.

THE APATITE OR PHOSPHATIC DEPOSITS OF NORWAY.

The existence of apatite in Norway has been known for a long time, and special attention was drawn to it by M. Dufreynoy in his 'Traité de Minéralogie.' The discovery of deposits which could be profitably worked only dates, however, from the year 1871. The discovery was made accidentally by Peter Simonsen, who organised a French company to work the deposits he had discovered. The explorations of this company led to the discovery of a number of similar deposits within the same district. The only mines at present worked extensively are those which are worked by a French company in the parish of Bamle, near Oedegaarden.

The region of the principal phosphatic or apatite deposits known in Norway stretches, as will be seen by a reference to the map, fig. 14, from the towns of Stathelle and Langesund, west-south-west, to the port of Kragerö, and on to Risor and the neighbourhood of Arendal, as shown in the map, fig. 14. I have made several visits to this region, and I have examined a great number of these deposits. I am enabled, therefore, to supplement the interesting information given in the article[1] referred to below from personal observation.

The section, fig. 15, will show the general order of the strata at this region, which belong to the Laurentian group,

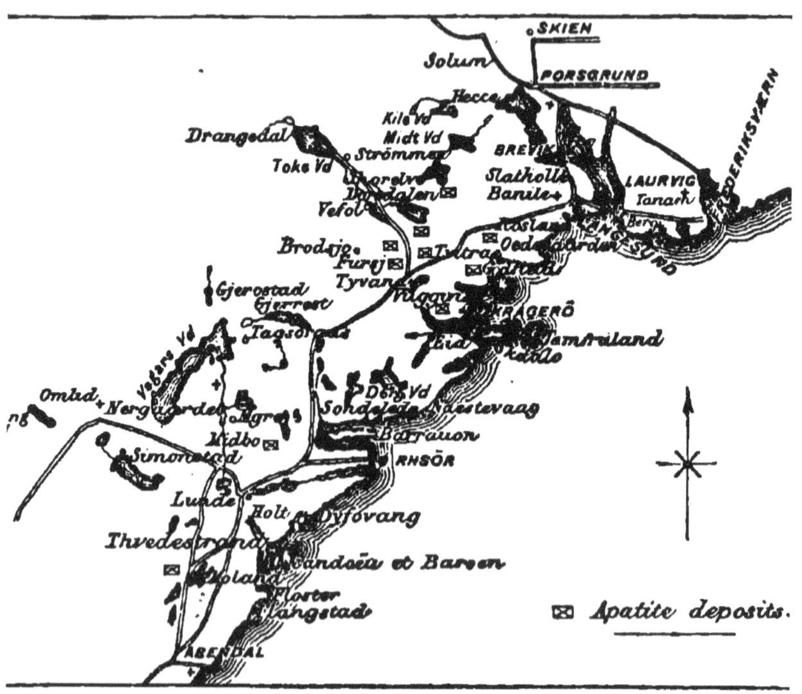

FIG. 14.—MAP OF THE APATITE DISTRICT, NORWAY. Scale, 8 English miles = 1 inch.

and bear a great resemblance to the strata of the same age in which the Canadian phosphatic deposits occur.

The rock in which the apatite is usually found is a dark grey granular gneissic rock, with a large proportion of hornblende in it, the colour being lighter or darker according as there is less or more of this mineral. The particles of the

1 'Vorkommen des Apatit in Norwegen,' Herren Brögger und Reusch. *Deutschen Geological Gesellschaft*, 1875, p. 646, *et seq.*

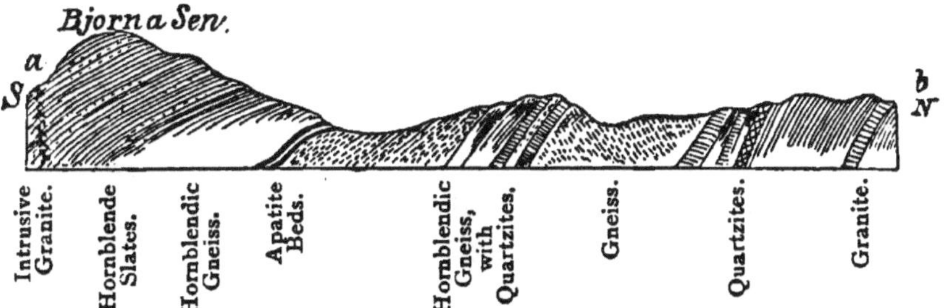

FIG. 15.—SECTION FROM AAS TO FOSTHI, SHOWING GENERAL ORDER OF STRATA IN APATITE DISTRICT, NORWAY.

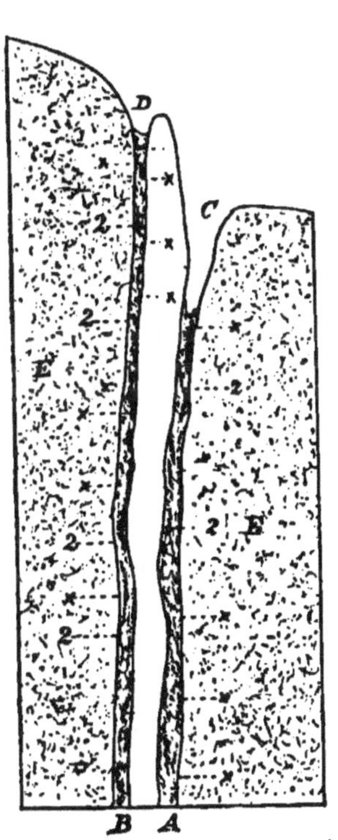

FIG. 16.—VEINS OF APATITE AT TVITRAE, NORWAY.

A B, Apatite Veins.
C D, Openings.
Dark shading, Hornblende.
Light parts with x, Apatite.
" " 2, Quartz Strings.
E E', Gneissic Rock.

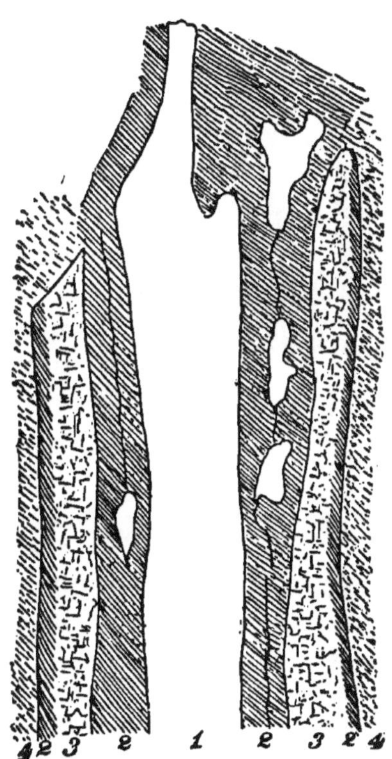

FIG. 17.—APATITE VEINS, GODFIELD.

1, Apatite.
2 2 2, Hornblende, large flakes of Mica, and some Titanic Iron Ore.
3 3, White and Grey Granular.
4 4, Dark and Grey Granular Rock.

different minerals the rock is composed of are less regularly arranged than in ordinary gneiss.

The apatite occurs in veins, as shown in figs. 16, 17, 18; in nests and pockets, as seen in figs. 19, 20, and 22—fig. 22 giving a fair illustration of the surroundings of a Norwegian apatite mine; and in beds, as shown in the sections, figs. 15, 23, and 24. The *Gabbro* referred to in the explanation of some of these figures is the name given to darker and rougher varieties of the hornblendic gneiss, in which the deposits are found. It also occurs in large crystals, fig. 22, groups of which sometimes take the place of the nests and pockets distributed through the rocks. The veins do not lie along lines of displacement of the strata, but are rather cracks of shrinkage filled with phosphate and its associated minerals. Fig. 16 represents two parallel lodes, about 8 ft. apart, at Tvitrae, which can be traced along the surface for two or three hundred yards, and the southern one has been followed to a depth of 30 ft. They are from 3 ft. 6 in. to 6 ft. wide, and the apatite, which here is chiefly red in colour, with a little yellow, lies in pockets and wedge-shaped masses within the lodes. These masses are sometimes 4 ft. thick and 6 or 8 ft. long, and they are connected by strings of apatite and hornblende, sometimes by the latter alone, the miner never losing heart as long as he has a string of hornblende to follow. The hornblende also surrounds the apatite as a black margin from ½ in. to 1 in. thick. These apatite masses are separated from each other by similar masses and strings of dull quartz, fig. 21, so that probably not more than two-fifths of the contents of the veins are

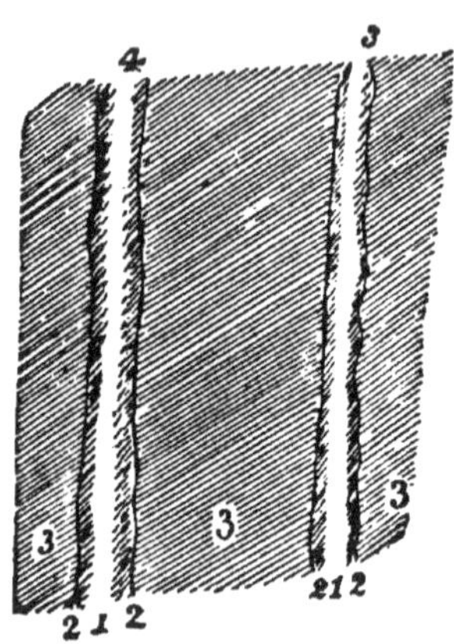

Fig. 18.—SECTION OF DOUBLE LODE, GODFIELD.

Scale ¼ inch = 1 foot.

1 1, Apatite with occasional strings of Quartz.
2 2, Hornblende (Black) with Iron Ore occasionally.
3 3, Grey Granular Rock.
4, Hornblendic Gneiss.

apatite. Figs. 17 and 18 are also illustrations of apatite veins as they occur amongst a number of others at Godfield, where the apatite is of a cream and greenish colour. Here are the usual conditions—the grey granular rock, the fringe of hornblende, which on each side of the apatite in fig. 17 contains some beautiful crystals of titaniferous iron ore and large flakes of mica. In this illustration are also seen examples of pockets running alongside the vein. Being simply cracks of shrinkage, these veins are uncertain as to their continuance in length and depth, and also as to the character of their contents. It is

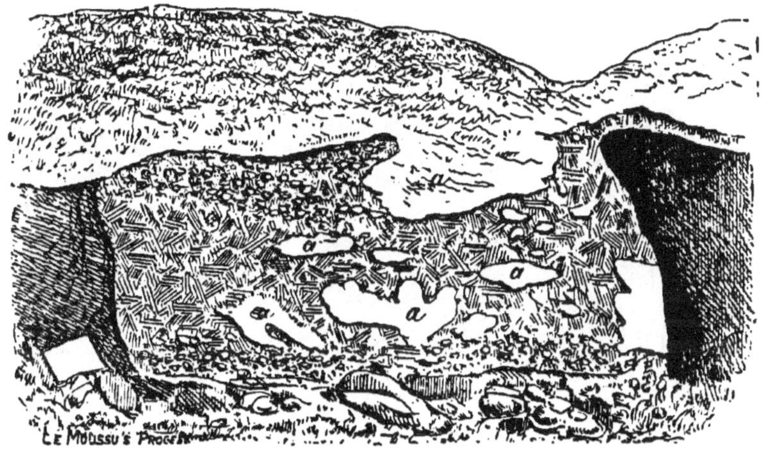

FIG. 19.—VUGGENS APATITE MINE NEAR KRAGERÖ, NORWAY, SHOWING NESTS OF APATITE, *a*, *a*.

only when a long length is exposed, as at Tvitrae, and at the works of the French company near Faesset, that their average worth can be estimated.

In figs. 19 and 20 we have illustrations of the way in which the mineral occurs in pockets, and it will be seen that the chief thing to be considered in estimating the commercial value of such deposits is the proportion of apatite contained in the mass of rock to be removed. These deposits are uncertain in their character. Fig. 20 is an illustration of the character of a long cluster of nests about 50 ft. in length, which promised to give an average thickness of 1 ft. of apatite;

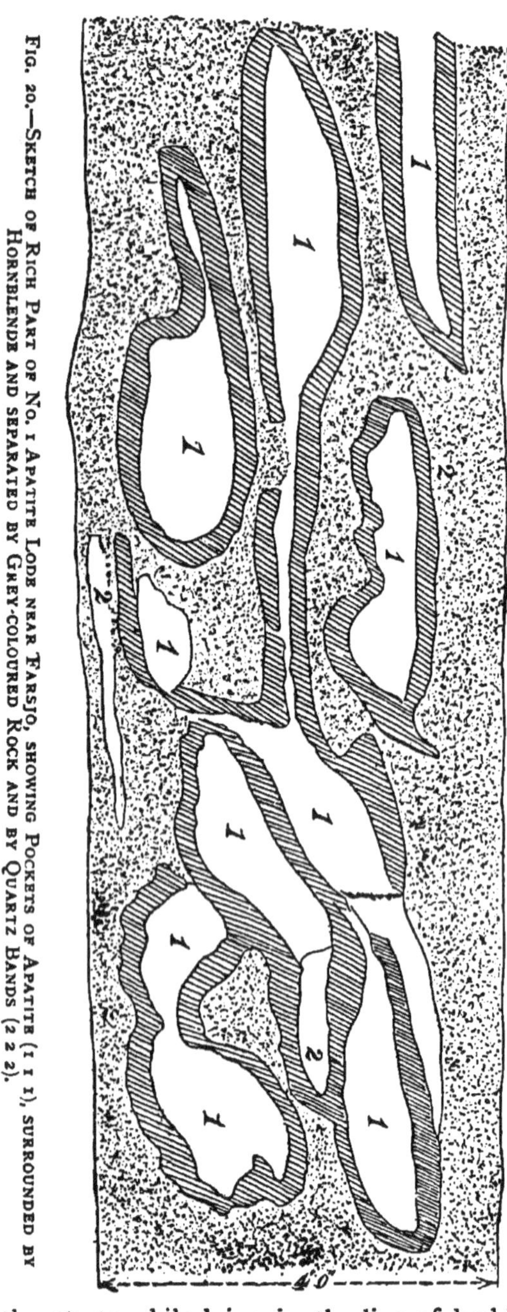

Fig. 20.—Sketch of Rich Part of No. 1 Apatite Lode near Farsjo, showing Pockets of Apatite (1 1 1), surrounded by Hornblende and separated by Grey-coloured Rock and by Quartz Bands (2 2 2).

but the whole of these nests died out at a depth of a few yards. The Vuggens mine, fig. 19, has been more successful. Here again the apatite is surrounded with hornblende, the crystals of which in all cases point in length towards the apatite.

Fig. 23 is an illustration of a series of bed-like deposits on Doredalen property, near the Tvitrae Lake. These beds crop out on the high breast of a hill, and are traceable for a considerable distance. There are six or seven of them, ranging from 6 in. to 2 ft. in thickness. Possibly, however, they may not be true beds, but collections of apatite occupying portions only of the strata while lying in the line of bedding. They are, however, the most massive and continuous beds I saw in Norway. The apatite here varies in colour from cream through green and white and pink. At a little distance from these beds there is a long irregular mass about 30 ft. long, lying in the line of the stratification and dying out to a string at each end. Another interesting series of bed-like deposits occur near the south-west end of the range in the hill Högaasen, on the Midbo property, about half-way between Tvedestrand and Arendal. Fig. 24, from a sketch taken by me in August, 1882, illustrates these deposits, which are interesting from the amount of rutile—oxide of titanium associated with the

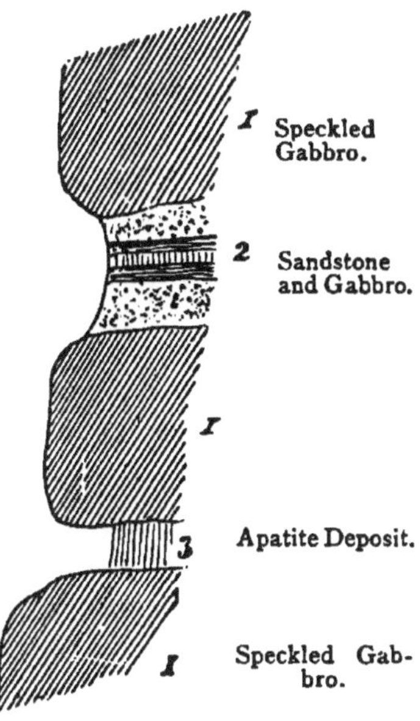

Fig. 21.—Section of Apatite Deposit at Oedegården.

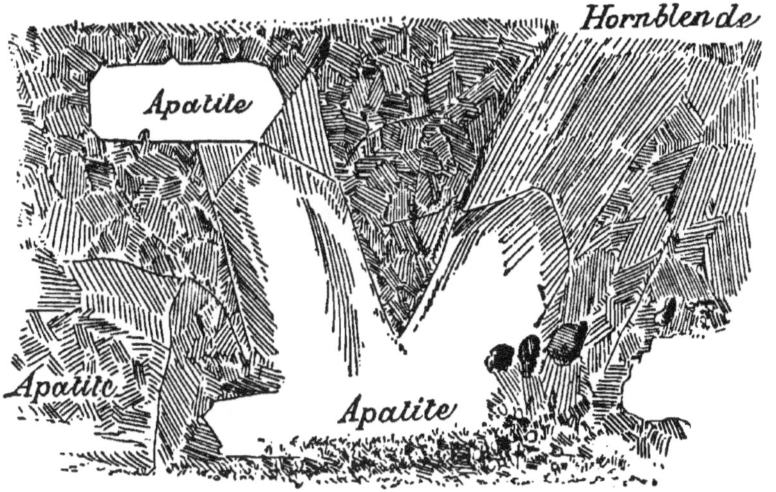

Fig. 22.—Fine Crystals of Apatite in Vuggens Mine, Kragerö, Norway.

apatite, as well as for the beautiful separate crystals of apatite.

The latter mineral lies in pockets, the run of which coincides with the bedding of the gneissic rocks.

The following analysis by Dr. Voelcker of two samples each of red and of white will give an idea of the rich quality of these Norwegian apatites.

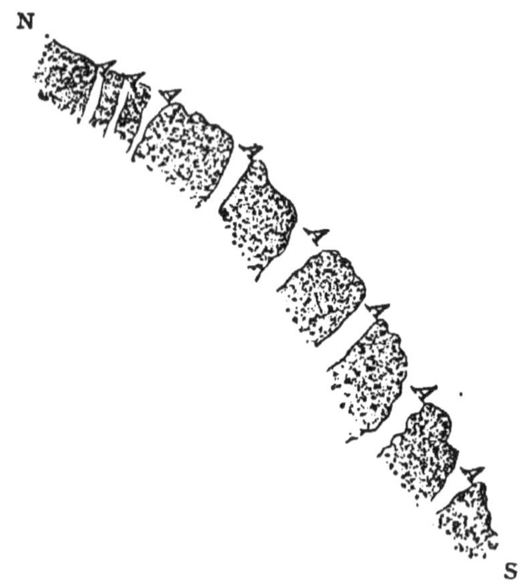

FIG. 23.—HILL-SIDE ON DOREDALEN PROPERTY, NORWAY, WITH APATITE BEDS (A A A) SET IN GREY HORNBLENDIC GNEISSIC ROCK.

	Red Apatite.		White Apatite.	
Hygroscopic water	0·43	0·43	0·19	0·298
Water of combination	0·40	0·40	0·23	0·198
Phosphoric acid	41·88	41·74	41·25	42·280
Lime	53·45	54·12	56·62	53·350
Chloride of calcium	1·61	1·61	6·41	2·160
Magnesium	—	0·20	0·29 }	0·920
Iron and alumina	1·66	0·45	0·38 }	
Insoluble matter	1·24	0·97	0·82	0·990
Alkalies	—	0·50	0·17	—
	100·67	100·00	100·36	100·196

From the absence of all traces of the remains of organic life in the apatites of Canada and Norway and in their surroundings, we may reasonably infer that in them we have original deposits of apatite from phosphatic matter disseminated in the water of those early seas, derived probably from gaseous emanations and eruptions from the interior of the earth, and deposited pure and simple, without having passed through the structure and substance of living organisms. The gathering of

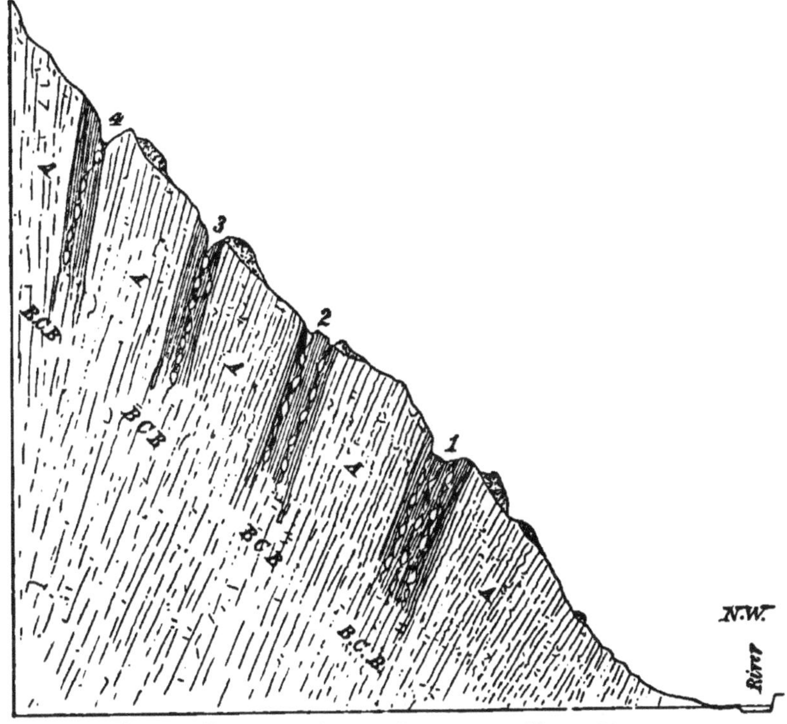

FIG. 24.—SECTION OF APATITE DEPOSITS AT MIDBO, NORWAY.

A A A A, Red and Grey Gneissic Strata.
B B B B, Dark-coloured Hornblendic Gneiss bounding the Apatite Courses C C C C.
1, 2, 3, 4, Openings on the Apatite Courses.

the mineral into separate masses distinct from the rest of the strata, with the crystalline fringe of hornblende and titaniferous iron ore, indicates considerable chemical action with its resulting crystalline conditions subsequent to these depositions of the phosphatic matter.

The containing gneissic rock is often varied by passing into large masses of pink and red felspar, especially in the immediate vicinity of the apatite pockets. Dykes of the same substance, and also of granite, not unfrequently cross the strata and cut off the veins. Illustrations of this occur at Godfield.

The chief element of uncertainty in the mining of these deposits lies in the proportions of apatite there may exist to the enclosing matrices, and this, of course, depends upon the size of the nests and bunches of apatite, and the distances at which they may lie apart.

A difficulty also occurs in the dressing, from the great similarity there is between the red felspar and the red apatite, and between the white apatite and the white quartz. It requires some degree of handling and of familiarity to the eye to distinguish the difference. It is also next to impossible in actual mining to secure the whole of the apatite. It is tender, and in the process of blasting it gets broken up. Even with the most careful screening and picking it is impossible to avoid having a second quality of the very small of from 45 to 50 per cent., and after all some portion is lost.

It will be seen that with deposits so uncertain in their character it is difficult to fix the costs of mining and dressing the mineral. It may be assumed that miner's wages are less in Norway than in England, but that this is balanced by the rather higher price of materials. Then it may be taken that an area or space of apatite 6 ft. by 6 ft. by 4 in. thickness is equal to a ton, and also that the whole of this in mining cannot be saved for use. Taking apatite of the quality of 85 per cent., and an average thickness over a considerable area of 8 in., and a production of from 1,000 to 1,500 tons a year, the cost of mining, dressing, and shipping a ton of apatite, including cost of management and all other costs, if the mine is within four or five miles of a port, may be taken at 65*s.* It is essential to success in apatite mining in Norway that a large number of concessions be grouped under one management, not only in order to save in cost of management, but also that when one mine is poor another may be rich, and the supply be kept up.

CHAPTER VIII.

PHOSPHATE OF LIME—continued.

Phosphatic Matter in Strata between the Laurentian and Lower Silurian—The Phosphorite Deposit of North Wales—Discovery—Range—Associated Strata—Description of Bed—Supposed Origin of Analyses—Compared with other Phosphates—Particulars and Costs of Mining and Dressing—The Phosphatic Deposits of Estramadura, Spain—Position—Discovery—Composition—Description of Particular Deposits—Deposits in Canada, France, Hungary—Bone Bed at Top of Upper Silurian Strata in Shropshire.

DEPOSITS IN SILURIAN STRATA.

BETWEEN the deposits of phosphate of lime described in the last two chapters and those now to be described there is a vast thickness of strata, probably not less than 70,000 to 80,000 feet in thickness, comprising the uppermost beds of the Laurentian group, the whole series of the Cambrian, and three-fourths at least of the Cambro-Silurian or Lower Silurian groups of strata.

Throughout the whole of this vast series of rocks there are traces of phosphatic matter, which becomes an appreciable quantity where organic remains abound, as in the Paradoxides beds of St. David's, the fossiliferous beds of the Lingula strata, and of the Llandeilo limestone.

At nearly the summit of the Cambro-Silurian strata in North Wales, and resting on the uppermost surface of the Bala limestone, we meet with a deposit of phosphate of lime, which is interesting scientifically, and which, although it has not yet been extensively worked, is, I think, from its position, quality, and the great extent of the deposit, destined to be.

Dr. Voelcker, who had been employed to make analyses of

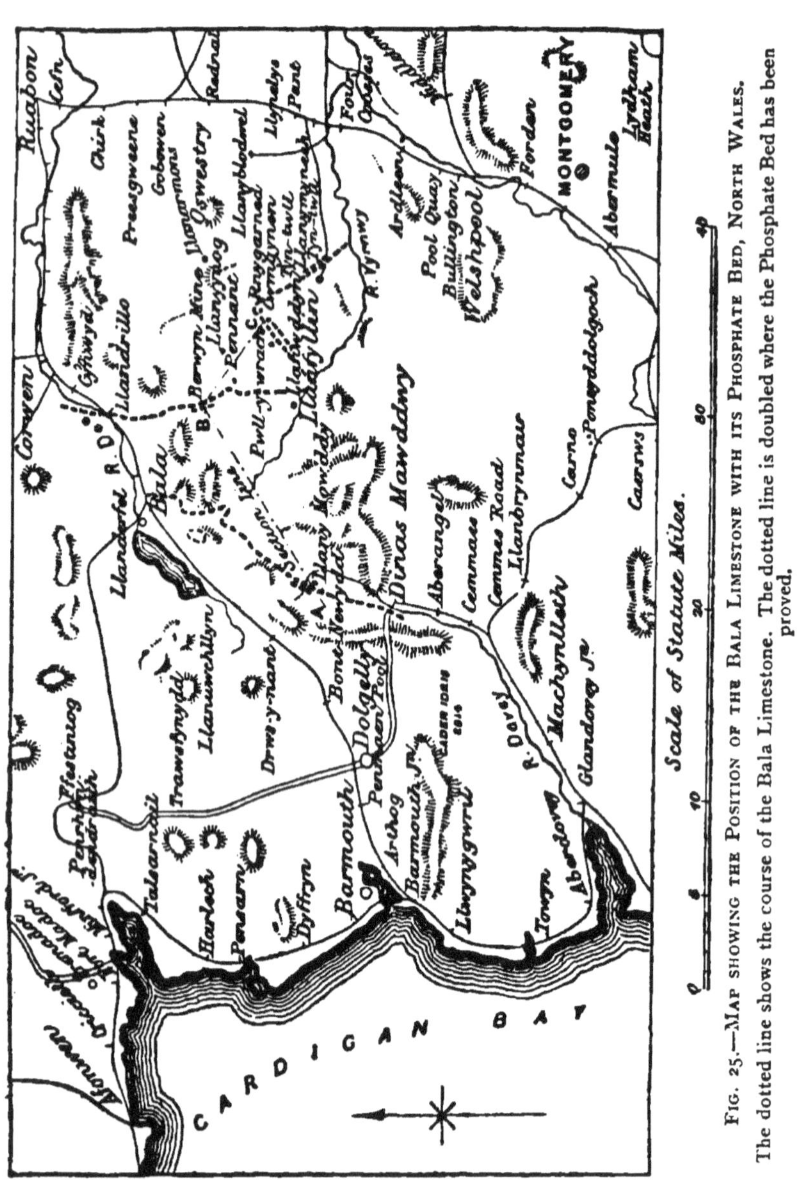

FIG. 25.—MAP SHOWING THE POSITION OF THE BALA LIMESTONE WITH ITS PHOSPHATE BED, NORTH WALES. The dotted line shows the course of the Bala Limestone. The dotted line is doubled where the Phosphate Bed has been proved.

some samples of the deposit for the original discoverers and

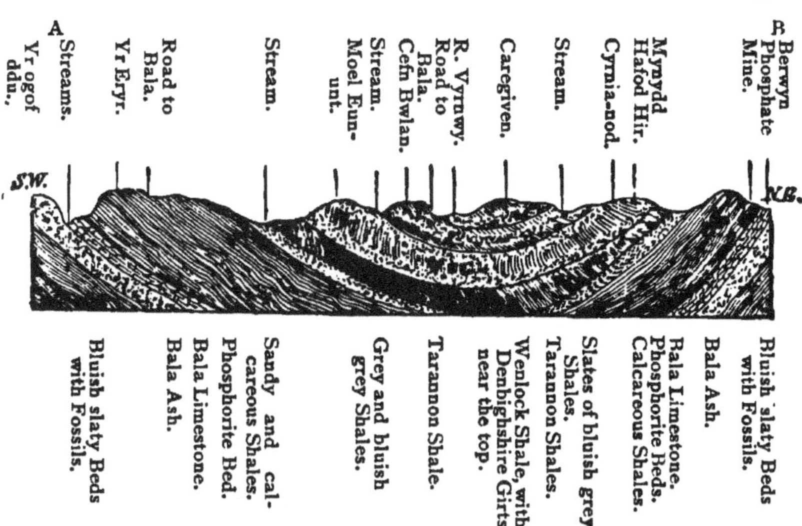

FIG. 26.—SECTION ACROSS THE BERWYN AND ARRAN MOUNTAINS, N. WALES, ON LINE A B OF PLAN. Scale ½″ = 1 mile.

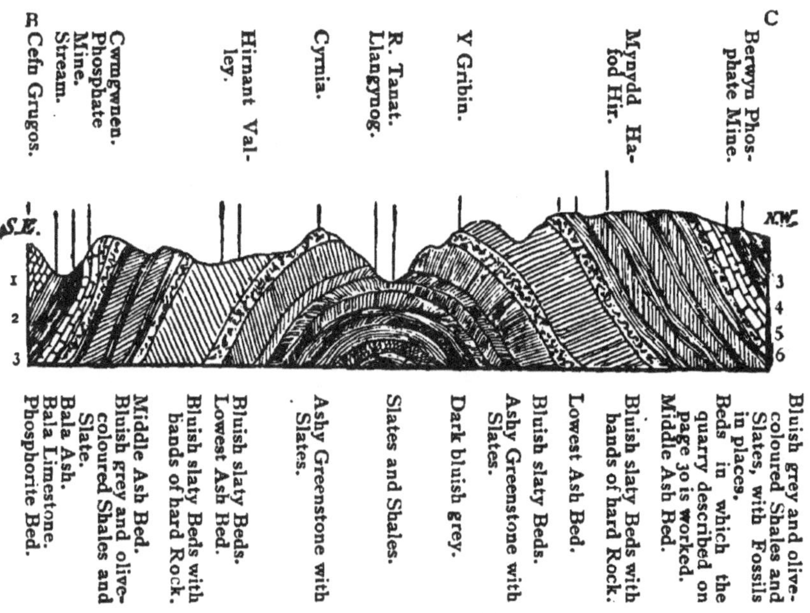

SECTION ACROSS THE BERWYN MOUNTAINS, N. WALES, ON LINE B C OF PLAN. Scale ½″ = 1 mile.

1, Hirnant Limestone. 2, Bluish Shales. 3 3, Shales with Phosphatic Nodules. 4, Phosphorite Bed. 5, Bala Limestone. 6, Bala Ash passing into Greenstone.

workers, first directed public attention to it at the meeting of the British Association held in Birmingham in the year 1865, when he entered into particulars concerning its composition and value. In the year 1867, having been previously employed in the examination of the deposit at points east and west of the place of original discovery, I communicated a paper to the *Geological Magazine*,[1] in which its true stratigraphical position and mineralogical conditions were first described.

Subsequently, in 1871, a discovery was made of a similar bed near the summit of the Berwyn Mountains, between Llangynog and Bala, with which I became connected, and having had occasion to examine the deposit at various other points of discovery, I embodied the whole of my observations in a communication made to the Geological Society in 1874, and which appeared in the journal of that society during the following year.[2]

I will here summarise those observations, and add such particulars of results and cost of working as will illustrate the value of the deposit, and which will, I hope, be of service to those who may hereafter attempt its working.

The deposit, as shown on the sketch-map, fig. 25, follows the course of the Bala limestone in the north-east part of the county of Montgomery, North Wales. The section, fig. 26, along the line A B C of the map, illustrates the general structure of the county, and shows the phosphate bed with its underlying limestone in the same position over the whole area.

Although not essential for our purpose, it will be interesting to note the detailed structure of the Bala limestone as it is seen at the Berwyn phosphate mine, inasmuch as probably it forms the most complete detailed section of that series of beds in North Wales.

[1] 'On a bed of Phosphate of Lime in North Wales.' D. C. Davies. *Geological Magazine*, vol. iv. p. 257.

[2] D. C. Davies 'On the Phosphatic Deposits of North Wales,' *Quarterly Journal Geol. Society*, vol. xxxi. p. 357.

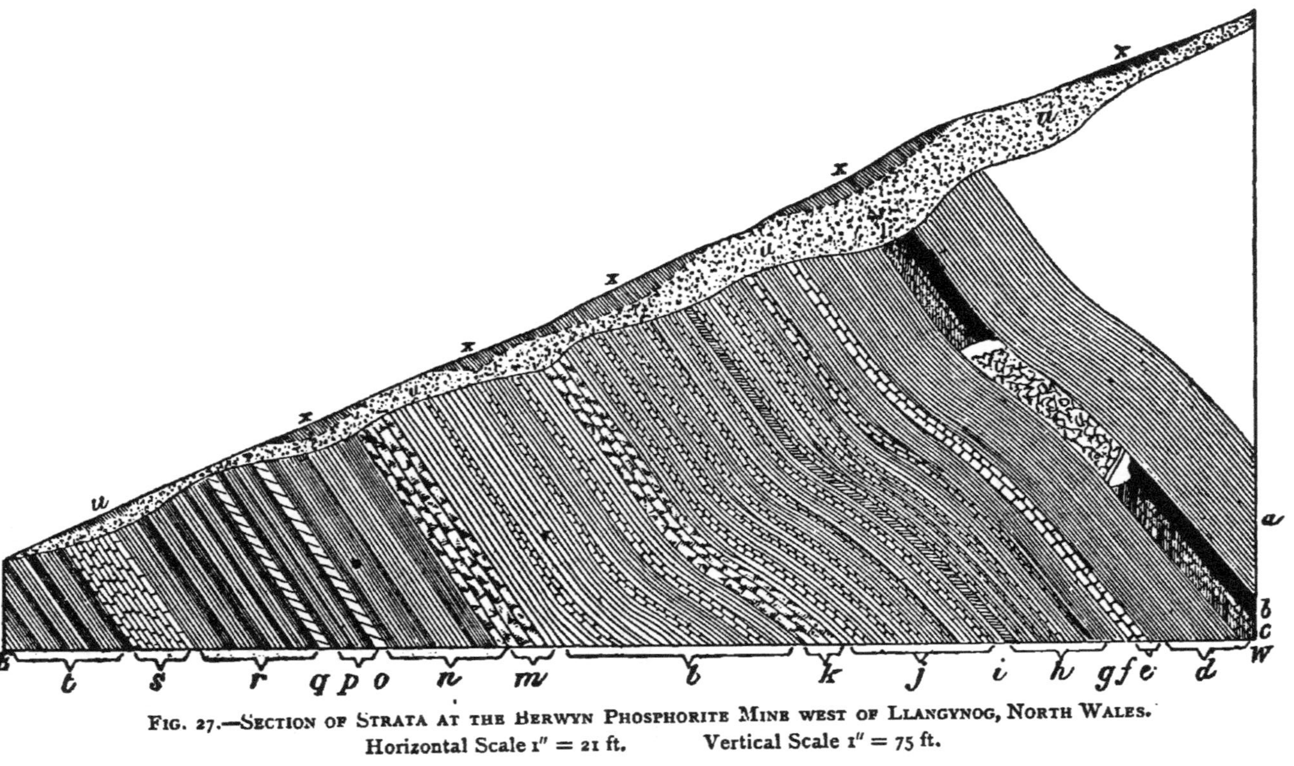

Fig. 27.—Section of Strata at the Berwyn Phosphorite Mine west of Llangynog, North Wales.
Horizontal Scale 1" = 21 ft. Vertical Scale 1" = 75 ft.

a Grey shale with echinoderms and other fossils phosphatised.
b PHOSPHATE BED.
c Dark limestone impregnated with phosphatic matter.
d Dark shales with veins of sulphate of baryta, passing upwards into a soapy clay.
e Limestone with veins of sulphate of baryta.
f Dark shales.
g Bluish grey limestones.
h Limestones and shales, often pyritous and decomposing towards the top as brownish sandstone, containing fossils *orthis* and *leptæna* of several species, also trilobites—illaenus and asaphus.
i Blue slaty bed.
j Limestones and shales with fossils, *orthis*, *leptæna*, bellerophon, &c.
k Crystalline limestone.
l Tough blue shaly rock with limestone partings.
m Limestones with the usual fossils, plentiful.
n Beds of bluish shale with black balls of phosphate and small trilobites.
o Kaolin.
p Tough calcareous shales, fossiliferous.
q Kaolin.
r Alternations of slaty, calcareous, and arenaceous beds.
s Limestone, composed almost entirely of *orthis spiriferoides*.
t Shaly beds slightly calcareous, with limestone bands.
u Superficial drift.
x Peaty deposits.

There is a great similarity of structure in many respects in this group of beds all over the district under consideration.

The phosphate bed lies at the top of the series of limestone beds, and is overlaid by the shales, *a*. It varies in thickness from 6 to 18 inches. It is black in colour from the graphite it contains; rarely, where the graphite is absent, it is of a pale yellow colour. The bed is made up of a number of concretions, which range in size from that of an egg to that of a full-sized cocoanut, which are closely packed together, and run into each other. They are cemented together by a black shaly matrix. The concretions have often a polished appearance, which is also due to the presence of graphite. Frequently, along the course of the bed, the phosphorite is charged with concretions and crystals of sulphide of iron.

Near the surface the pyrites become oxidised, and the deposit changes its black for a rusty appearance.

The concretions contain from 60 to 69 per cent. of phosphate of lime; but the matrix also contains a portion, so that the average quality of 1,000 tons sent off from this mine has been about 46 per cent.

Latterly, by more careful dressing and selection, the quality has been brought up to 55 per cent., and there need be no difficulty in sustaining this average. With proper appliances and means for drying the phosphate, the percentage may be increased to 58 or 60.

The bed is underlain by a thin bed of crystalline limestone, *c*, which does not usually exceed 6 inches in thickness, although at times it does thicken out and form a solid limestone 2 feet thick. This limestone contains phosphate of lime to the extent of 15 to 20 per cent.

Sometimes the phosphate bed is seen to divide into two, and more rarely, at the Berwyn mine, fig. 27, into three beds. When this division takes place the dividing substance is the phosphatic limestone. The uppermost bed dies out as it enters the shales; so does the middle one. It is invariably the lowest bed which is continued forward, the overlying limestone dying out until the shales take their true position above the phosphate bed.

In attempting to account for the existence of this bed we cannot be far wrong, I think, in ascribing to it an organic origin. It is probably an old sea bottom, on which the phosphatic matter of crustacean and molluscan life was precipitated and stored during a long period, while certain marine plants may also have contributed their share of phosphatic matter. There may also, as in the case of the Laurentian deposits, have been an abundance of phosphatic matter in the water of this early sea, independent of the living organisms which it sustained.

If we could bend down the edges of the strata of the section, fig. 26, to a horizontal line, and piece them with the middle portion, which has been broken and denuded through the upheaval of the underlying porphyries, greenstones, and slates; if we further follow the phosphorite bed underground along the section, fig 26, to where it comes up in altered form on the flanks of Aran Mawddy, and measure the length of the district described on the map by the breadth thereof, we should gain some conception of the extent of the shallow sea with its swarms of life in which the bed was deposited, covering as it does an area of about 150 square miles. Over this the depth of the sea must have been nearly uniform, and the same conditions of life must have prevailed.

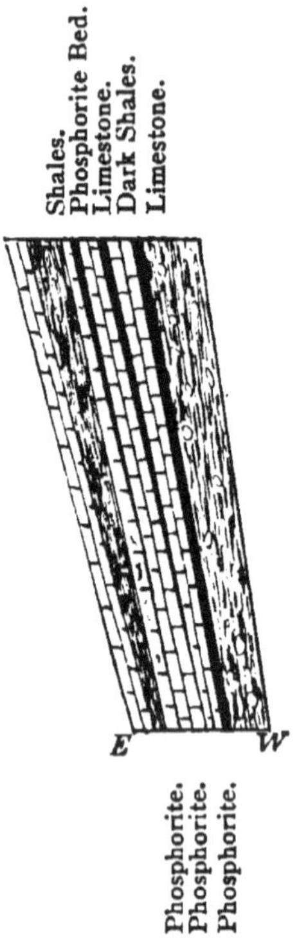

Shales. Phosphorite Bed. Limestone. Dark Shales. Limestone.

E W

Phosphorite. Phosphorite. Phosphorite.

FIG. 28.—SECTION SHOWING DIVERGENCE OF THE PHOSPHORITE BEDS AT THE BERWYN MINE.

Then I do not doubt that, in some shape or other, the bed may be found at the same horizon all along the course of the Bala limestone in North Wales.

The bed presents the same general appearance at all the points where it has been opened upon. To the east, about Llanfyllin, however, it becomes more sandy and impure, while on its western outcrop, about Llan-y-Mawddy, the phosphatic matter is largely replaced by sulphur. The figs. 29, 30, 31, 32, 33 will illustrate the details of the bed and its associated strata at widely different places.

The following analyses and particulars will show the character and commercial value of the deposit.

Analysis of nodules:—

Moisture and organic matter	4·200
Insoluble matter	22·600
Tribasic phosphate of lime	64·165
Oxide of iron and alumina	6·890
Other constituents not determined . .	2·145
	100·000

Another analysis of nodules gave 69·24 phosphate of lime, without any pyrites or carbonic acid, and only slight traces of sulphuric acid. Five analyses from the bulk of consignments from the Berwyn mine previous to 1876 gave an average of 46·85 of phosphate of lime.

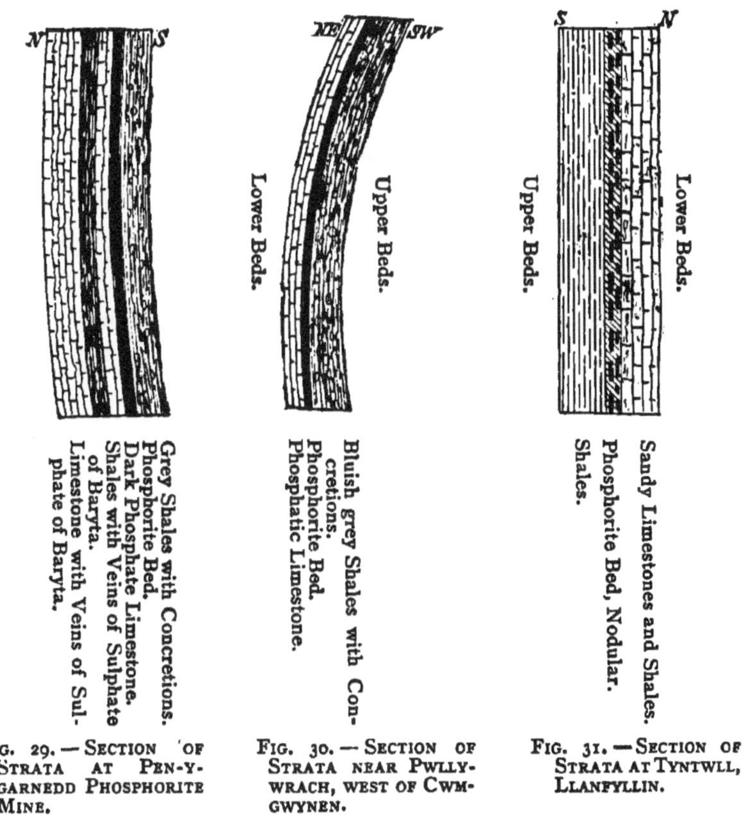

FIG. 29.—SECTION OF STRATA AT PEN-Y-GARNEDD PHOSPHORITE MINE.

FIG. 30.—SECTION OF STRATA NEAR PWLLY-WRACH, WEST OF CWM-GWYNEN.

FIG. 31.—SECTION OF STRATA AT TYNTWLL, LLANFYLLIN.

Dr. Voelcker's original analysis of two samples from Cwm-gwynen, the place where the bed was first recognised, were—

1st sample gave	64·16	of phosphate of	lime.
2nd „	48·50	„	„

There was no carbonate of lime, some fluoride of calcium, alumina, and oxide of iron.

The darker-coloured contained more graphite and were richer in phosphate of lime than the light-coloured specimens.

The deposit at Pen-y-garnedd, when properly dressed, averaged 46 per cent. of phosphate of lime.

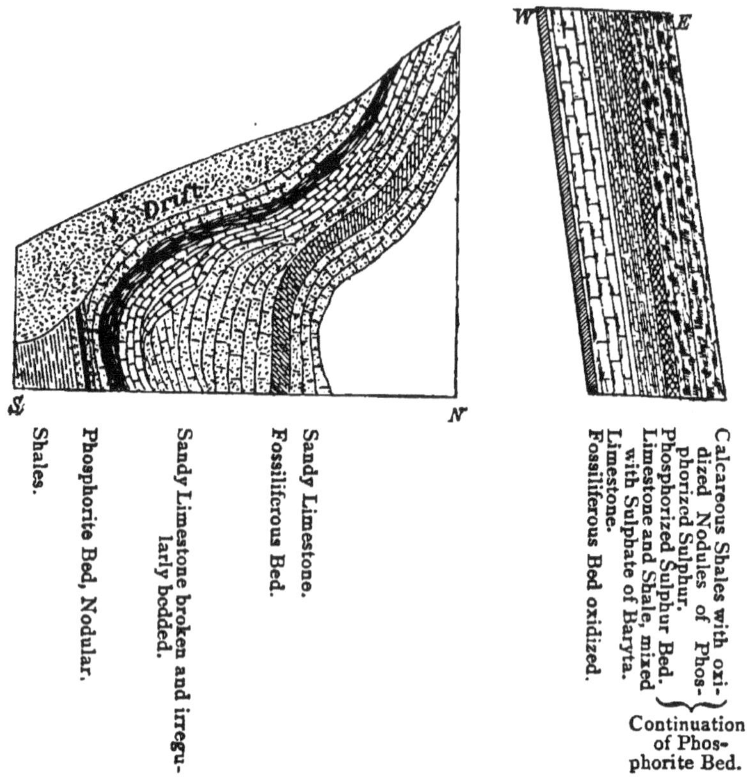

FIG. 32.—SECTION OF STRATA AT GREEN HALL PARK, LLANFYLLIN.

FIG. 33.—SECTION OF STRATA AT CWM-DYNEWYDD, NEAR LLAN-Y-MAWDDY.

Two samples from near Llan-y-Mawddy, where the phosphate is replaced by sulphur, gave the following results :—

Phosphate of lime . .	2·90	1·72
Sulphur	34·38	34·20

Partly through want of knowledge, and partly through carelessness, some of the early consignments of phosphate from Cwmgwnen and Pen-y-garnedd, and later from Berwyn, were not properly dressed and selected. The result was a percentage as low as 30, and the Welsh phosphates came into disrepute.

The following recent analyses from consignments in bulk, and which were made by the analysts of the consignees, will show what these Cambro-Silurian phosphates can really be made to do, while, as I have said, with proper appliances for dressing and means to dry the mineral, it may be sent off 4 or 5 per cent. higher in quality.

1. Analysis of sample of 5 tons, by D. H. Richards, F.C.S., borough analyst of Oswestry :—

Insoluble in hydrochloric acid	31·020
Water	280
Lime	32·256
* Phosphoric acid	23·572
Oxide of iron, &c.	3·686
Carbonic acid	2·220
Not determined (alkalies, &c.)	6·966
	100·000

* Equal to tribasic phosphate of lime . 51·46

2. Analysis and samples of 8 tons, by Dr. Voelcker :—

Water and loss on heating	5·49
* Phosphoric acid	23·25
Lime	32·70
Oxide of iron, alumina, magnesia, carbonic acid, &c.	15·04
Insoluble silicious matter	23·52
	100·00

* Equal to phosphate of lime . . 50·76

3. Analysis of sample of 8 tons, by Nesbitt & Co. :—

Moisture	0·35
Water of combination	2·65
Silicious matter	28·75
Oxide of iron and alumina	4·60
Lime	32·62
* Phosphoric acid	23·79
† Carbonic acid	2·80
Sulphuric acid	0·96
Undetermined	3·48
	100·00

* Equal to phosphate of lime . . 51·95
† „ „ carbonate of lime . . 6·36

Two subsequent analyses, one of 8 tons, by E. Davies, F.S.C., of Liverpool, gave 54·97; and one by the analysis of Messrs. Newton Keats & Co., of St. Helen's, gave 55·80.

The value of phosphates of lime for agricultural manures depends largely upon the absence of oxide of iron and alumina. If these substances are present in quantity the phosphate, after it has been made soluble, becomes fixed again, or goes back to an insoluble state. Further, if there is too much carbonate of lime, the quantity of sulphuric acid required is very great, owing to the acid attacking the carbonate instead of the phosphate of lime. I have therefore been at some pains, in order to arrive at the true value of this North Wales deposit, to compare it in this respect with others of similar strength from various parts of the world. The annexed table is the result; and it will be seen that the phosphate from North Wales is really superior in these particulars to the others with which it is compared.

In mining the deposit a level is first driven along the bed from the hillside. An opening is next made up to the surface; then the bed is taken down by overhand stoping. It is found necessary to take from three to five feet of the limestone bed below the deposit away first. This is done by blasting; and if strong explosives are used, and care is taken to compel the miners to bore deep holes far apart, and to put strong charges of the explosive used, this operation may in future be done more cheaply than in the past. The phosphate bed is left standing for several fathoms in length, and when a sufficient area of its under surface has been laid bare, a few shots put in between the bed and the overlying shales brings down the whole mass, the parting between the bed and the shales being very distinct and clean. The whole of the bed obtained is taken out of the mine to be dressed, while the limestone is used to fill up the space mined below, passes being left at frequent intervals for the purpose of throwing the phosphate down into the level below.

The average amount of ground stoped by two men per week at the Berwyn mine was, in 1876, 47 square feet forward, or a little over 1⅛ fathom. With the use of stronger explosives, as I have said, the amount should be brought up to 2 fathoms.

The average cost for fuze, powder, and candles per pair of men per week was 4*s*. 8*d*.

The average nett wages made by men at 36*s*. per fathom, the men finding their own stores, was 23*s*. per week.

The average production of the bed over a space of 360 fathoms was 2 tons 10 cwt. of phosphate per fathom, of an average strength of 46 per cent.

The average yield of 53 per cent. quality from the bed was, as nearly as I can ascertain, 2 tons per fathom, the remainder, although containing a good deal of phosphatic matter, being rejected.

The cost of driving the level was from 4*l*. 10*s*. to 5*l*. per fathom.

The cost per ton of dressing 586 tons of ore up to an average percentage of 46 was 3*s*. per ton.

The Amounts of Oxide of Iron, Carbonate of Lime, and Alumina in various Samples of Phosphates yielding under 57 per cent. of Phosphate of Lime.

—	Oxide of Iron.	Alumina.	Carbonate of Lime.	Insoluble Sulphate of Iron, innocuous.	Soluble Sulphur and Fluorine.	Silicious Matter.	Phosphate of Lime.
Boulogne coprolites, ordinary quality .	6·24	5·39	8·95	..	3·24	26·16	45·19
Ditto, superior . .	14·38 (oxide of iron and alumina)		..	..	..	17·62	54·79
Cambridge, ditto . .	18·70 (oxide of iron and alumina)		..	..	..	7·77	56·87
Bedford, ditto . . .	5·29	7·24	7·84	..	..	20·81	51·24
Spanish Phosphate .	14·0 (oxide of iron and alumina)		..	..	..	37·21	56·19
German, ditto . . .	5·6	10·32 (alumina and carbonate of lime)		..	..	17·80	56·80
Carolina, ditto . .	3·99	3·20	6·61	..	..	not given	52·72
Welsh, ditto . . .	1·01	1·06	9·0	7·0	..	24·81	50·08

Average amount of oxide of iron, carbonate of lime, and alumina in seven of the above samples . . . 16·61
Ditto, in the Welsh sample . . . . 11·07
Only about 1 per cent. each of oxide of iron and alumina.

There are, without exaggeration, millions of tons of this deposit in Montgomeryshire, which, when the prejudices existing against its colour and those created by carelessness or ignorance in the early days of its mining, are overcome, may be brought into the market at a fair profit to those who may undertake its exploitation.

THE PHOSPHATIC DEPOSITS OF ESTRAMADURA, SPAIN.[1]

These important deposits extend from Logrosan by Montanches to Caceres, following the line of railway which now extends from Estramadura to Portugal. Attention was first drawn to them in the thirteenth century by Bowles, an Englishman, in a description he then gave of the natural riches of the country, and in which he gave the name of phosphorite to the

[1] *Quarterly Journal, Geological Society*, vol. i. p. 52. *Ann. des Mines*, t. v., 1834, p. 175.

mineral, from its property of giving light in the dark when thrown upon burning wood. The deposits were referred to by Le Play, in 1834, but they were first fully examined and described in 1845 by Dr. Daubeny and Captain Widdrington, who went into the inquiry as to whether the mineral would pay for working and for transport to England.

The deposits consist of a series of beds intercalated between schisty and slaty strata of probably Silurian age. The direction is N. 45° E., and they dip down almost vertically 70° S.E. There is a granitic bed not far below their horizon in the strata. At Logrosan these beds are about 40 yards across, and the single beds are sometimes worked in open trenches to depths of from 25 to 50 yards.

The beds have been regularly worked since the year 1865, and the phosphates imported into Great Britain have been estimated to reach on an average as much as 200,000 tons a year.

The deposit contains a maximum quantity of 85 per cent. of phosphate of lime near Logrosan and Montanches, and a minimum of 50 per cent. near Caceres. The beds are traceable for long distances on the surface. They differ from each other, and each one differs in composition and structure in different parts of its course. In all of them there is the presence of carbonate of lime, which in a certain measure forms a guide for the discovery of the phosphorite itself. There is iron in a quite large enough proportion, as well as of silicic acid. These phosphorites vary from a white to a light ochrey colour. The following is an analysis by Dr. Daubeny of phosphorite from Logrosan.

Phosphate of lime	81·15
Fluoride of calcium	14·00
Peroxide of iron	3·14
Silica	1·70
	99·99

The following particular description of the deposits at various points will give a further idea of their character.

1. *Jingal.*—Bed recognised near a mill in the village of Jingal, as it is entered from Truxello. Its thickness is about 4 ft. 6 in. It is followed for a length of about 420 yards, and is seen at a still greater distance.

2. *Del Casillon.*—This bed passes under the church of Logrosan, and presents at the gate of the village a mass of phosphate 25 feet in thickness, and very pure. For a long distance the average thickness of the bed is about 6 feet.

3. *Nostra Senora del Consuelo.*—A bed cropping out on the side of the hill of that name. It is split up into thin beds which probably join in depth.

4. *Costanaza.*—This bed has been followed for a distance of more than 2 miles. It passes downwards like the others in the Sierra Boyales, and reaches the base of the Sierra de Custova, which rises above the village of Logrosan.

5. *Terras Colorado.*—A bed 2 yards thick, and proved for a length of 105 yards, and parallel to the next bed.

6. *Cumbre Bojera.*—Which is an important bed.

There are no organic remains in these deposits. It has, however, been argued that the presence of carbonate of lime indicates an organic origin, and that traces of organic life may have been destroyed by the heat evolved in the irruption of supposed igneous rocks close by. To this it is sufficient to answer that there must have been both carbonate and phosphate of lime present in those early seas before organic life could exist, and that these substances represent the cause as well as the effect of organic life.

Similar deposits to those described extend into Portugal. They lie, like those of Spanish Estramadura, in schists above granitic rocks, and their quality ranges from 65 to 70 per cent.

While phosphate of lime is not known to occur in workable quantities elsewhere in strata of Silurian age, in FRANCE nodules of phosphate are found in the *calymene* beds of the slates of Angers.

In HUNGARY similar nodules are found in strata of the same age, and in CANADA such nodules are found associated with the Lingulae of the Lingula flags at the base of the Silurian rocks. Spherical concretions of a brown or black colour, containing a good deal of phosphoric acid, are found in Galicia; also in the government of St. Petersburg, and in that of Novgorod, in rocks of similar age.

At the top of the Silurian strata in Shropshire, and just below the lowest beds of the Devonian, is a highly phosphatic bed, and one that is very interesting scientifically, because it contains the oldest known remains of vertebrate life. The bones it contains are associated with crustacean remains, but hitherto the bed has not been commercially workable.[1]

A similar bed is also found at the top of the Devonian strata at their junction with those of the Carboniferous Limestone. Another bed occurs at the top of the Trias strata, between them and the beds of the Lias. In this bed there are, in addition to fish and crustacean remains, the bones and exuvia of the huge reptiles which had at this stage appeared in the order and succession of life.

There are two bone beds still higher—one at the junction of the Lias strata with those of the overlying Oolite, and one at the summit of the 'Wealden,' and at the base of the Lower Greensand. None of these various bone beds have as yet been found of sufficient thickness to pay for working; but I mention them here as interesting from their position at the junction of several consecutive groups of strata, and also as forming a source from whence may have been derived, by disintegration and denudation, some of the rolled and rounded nodules, usually known as coprolites, which we shall have to notice as occurring in some of the overlying groups of strata.

[1] See papers by the author in *Leisure Hour*, October, 1877, p. 685, on 'Fertilizers and Food Producers.'

CHAPTER IX.

PHOSPHATE OF LIME—continued.

The Greensand and Gault—Position of Bedfordshire and Cambridge Phosphate Beds—Localities—Description of Beds with Fossil Contents—Analyses—Composition of the Phosphate Bed—Derivation of the Phosphate Matter—The Phosphatic Nodules of Suffolk—Conditions of taking Phosphate Lands—Phosphate Digging—Statistics of Productions—Phosphatic Deposits of the Ardennes and the Meuse in France and Belgium—Date of Discovery—Geological Position—Extent—Characteristics of the Deposits—Phosphate Deposits of Bellegarde, France—Geological Position and Fossil and Mineral Characteristics—Analyses—Phosphatic Deposits of the Cretaceous Strata of Russia—History of the Discovery of Mineralogical Features—Analyses.

THE PHOSPHATIC DEPOSITS OF THE WEALDEN AND CRETACEOUS STRATA.[1]

BETWEEN the summit of the Oolitic strata and the base of the massive beds of the chalk there is interposed a series of beds of sand and clay which are known as Greensand. These beds are subdivided into Lower Greensand, Gault, and Upper Greensand. In the counties of Bedford and Cambridge each of these subdivisions contains a bed of phosphatic nodules, the precise position of which is shown in the section, fig. 34. The lowest and perhaps most important of these beds has been extensively worked in the neighbourhood of Sandy, and to the

[1] P. B. Brodie, F.G.S., *On a Deposit of Phosphatic Nodules on the Lower Greensand of Sandy, Bedfordshire.* J. J. Harris Teall, *The Potton and Wicken Phosphate Deposits.* Cambridge: Deighton, Bell & Co., 1875. 'Rock of the Cambridge Greensand,' Harry Seeley, F.S.A., *Geological Magazine*, 1866, p. 302. 'On the Phosphatic Nodules of Cambridgeshire,' by the Rev. O. Fisher, F.G.S., *Quarterly Journal, Geological Society*, vol. xxix. p. 52.

north-east at Wicken and Potton, Bedfordshire. From the

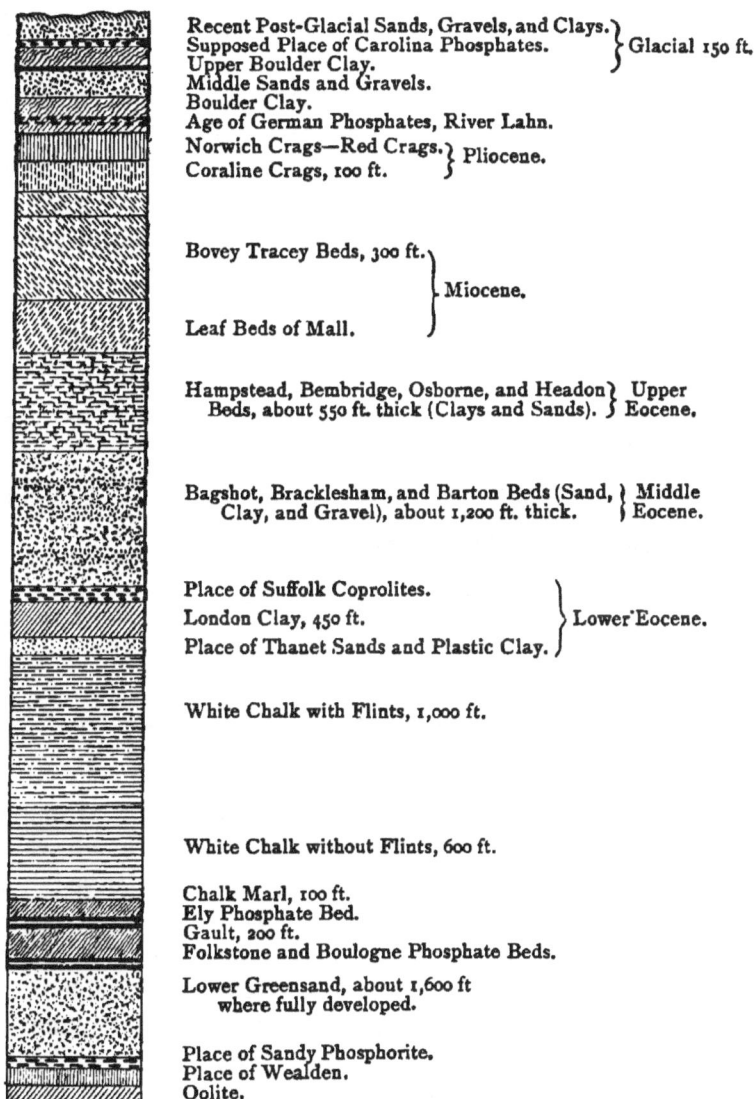

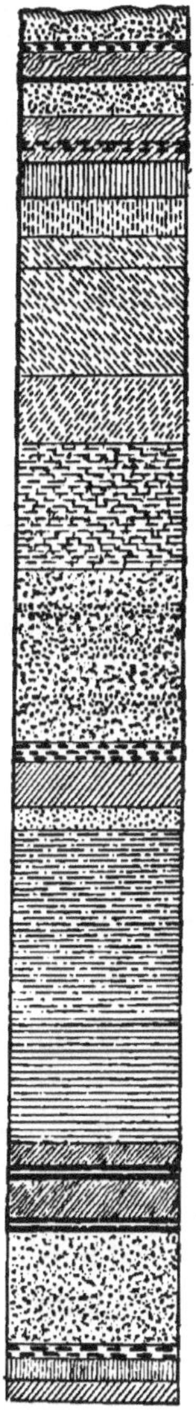

FIG. 34.—GENERAL SECTION OF STRATA FROM THE SUMMIT OF THE OOLITE UPWARDS, SHOWING THE STRATIFIED POSITIONS OF THE VARIOUS PHOSPHATIC DEPOSITS IN THE CRETACEOUS AND TERTIARY BEDS.

section fig. 35 it will be seen that the bed lies at the summit of sands about 50 feet thick, and that it is covered with similar sands and ferruginous sandstone. The bed varies in thickness from 6 inches at Potton to 2 feet at Sandy Heath. The bed is composed of phosphatic nodules and of pebbles in about equal proportions, and these are cemented together by ferruginous sand into a hard conglomerate. The nodules do not occur uniformly over a large area, but appear to run in patches, being occasionally absent. They vary in size from a pea to a hen's egg. They are of all shapes, rounded and elongated, are frequently pitted with minute holes on the outside, where they are of a light brown colour, but the interior is of a dark brown

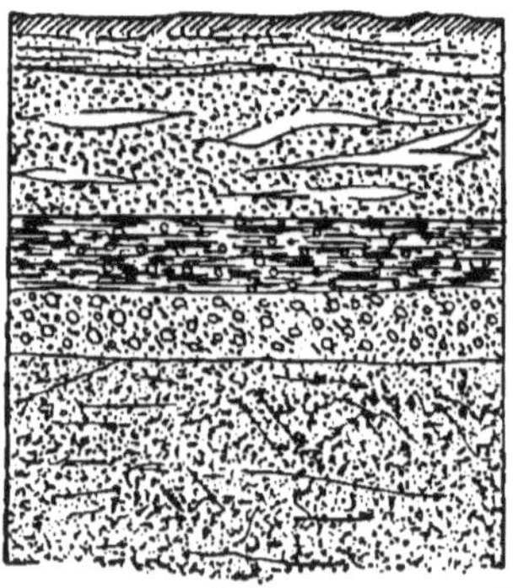

FIG. 35.—SECTION OF PHOSPHATE BED ON SANDY HEATH, BEDFORDSHIRE.

or black, and they often, but not always, enclose organic remains, casts, and fragments of fossils, chiefly those of ammonites. The enclosed and associated fossils are much waterworn. Of these fossils some—about half—are derived from other strata, about ten species of mollusca from the Portland beds, a large number of bones, teeth, and scales of the reptiles—iguanodon and megalosaurus—from the Wealden and Purbeck beds, with seven species of mollusca and numerous teeth and spines of fish from the Kimmeridge clay. The extraneous shells are often so rolled and broken as to defy recognition. Then about eighteen species of mollusca have been identified as belonging to the bed itself, inasmuch as they are not rolled or broken. Vegetable remains, including those of *Clatharia Lyelli*, are also found associated with the phosphatic nodules. The following analysis of the samples referred to by Mr. Brodie will show the composition of the nodules.

	Average Samples of Siftings, from layers at 1 and 2 ft.	Washed Coprolite from another spot.
Water of combination	5·17	5·67
Phosphoric acid*	22·39	15·12
Lime	32·73	26·69
Magnesia, alumina, and fluorine	6·64	4·51
Carbonic acid†	3·06	2·18
Oxide of iron	8·08	20·61
Silicious matter	21·93	25·22
	100·00	100·00
*Equal to tribasic phosphate of lime	48·51	32·76
†Equal to carbonate of lime	6·95	4·95

In the course of working these deposits care has been taken to bring the percentage of phosphate of lime as near as possible to the highest of these figures, the lowest being commercially valueless.

To the north of Cambridge, and near Ely, a similar bed, from six inches to a foot thick, occurs in what would seem to be a little higher place in the series of strata, resting as it does immediately upon the Gault. Thus nodules of this bed are richer in phosphatic matter than are those just described, a very full analysis by Dr. Voelcker giving the following results:—

Moisture and organic matter	4·68
Lime	43·21
Magnesia	1·12
Oxide of iron	2·46
Alumina	1·36
Phosphoric acid	25·29
Carbonic acid	6·66
Sulphuric acid	0·76
Chloride of sodium	0·09
Potash	0·32
Soda	0·50
Insoluble silicious matter	8·64
Fluorine and loss	4·96
	100·05

The greater part of the nodules seem to have been derived from other and older strata. Many of them show the structure and markings of the interior of shells of various kinds. All the derived fossils have plicatulæ attached to them, and even where they are broken plicatulæ are attached to the broken surfaces. They vary in size up to a diameter of four inches. Often they are irregular concretions, but frequently occur as tubes or halves of tubes. Although of a greenish cast outside, when broken they show a dark brown colour, and sometimes contain scales of fishes and small shells.

There are also fragments of bones of birds, reptiles, and fishes, all charged with phosphatic matter. As I have before hinted, some of these nodules may have been derived from the abrasion of older phosphatic beds, whether massive or nodular. There cannot, however, be any doubt that the water of this chalk sea was highly charged with phosphatic matter derived first from original sources, as the deposits of Norway and Canada, then passing into seaweed, bones, shells—horny and calcareous, and preponderating perhaps in the softer substance of the organisms themselves, and for a long time permeating all substances receptive of it, and gathering itself into various shapes around organic centres.

Higher up in the series, see section fig. 34, at the summit of the chalk and just above the London clay, we find the bed of phosphatic nodules worked in Suffolk. The nodules of this bed seem to have been derived in part from the London clay, and are due also in part to the operation of the causes already referred to. These nodules are not so valuable in commerce as those from Cambridge, containing as they do less phosphate of lime, more iron, and being of greater hardness. It was, however, the recognition of the phosphatic nature of these nodules by Professor Henslow, who regarded them as the exuviæ of extinct animals, that led to the discovering and working of the deposits already described.

In working these deposits in the three counties of Bedford, Cambridge, and Suffolk, the contractor pays the owner of the soil from 100*l.* to 140*l.* per acre. The average yield is about 300 tons per acre, and the value—until recently—of the nodules about 50*s*. per ton. When it is considered that the digger has to turn over from 3 to 15, sometimes 20 feet of overlying soil and sand, and to restore the land to its original condition, that he has also to wash, sort, and convey the nodules to a railway, it will be plain that the price paid to the landowner is too much, and that in the face of phosphates more cheaply won abroad, English phosphate-digging can hardly be profitably carried on. The fact, as shown by tables given further on, is that this branch of English industry has declined during the last seven years, and must finally die out if the present low prices prevail and such exorbitant royalty dues are charged. This will be especially true when the outcrop of these beds is exhausted and the nodules, if won at all, will have to be followed in depth by mining. The outcrop of the Greensand and Gault strata may be followed southward to Folkestone, and they contain more or less nodular phosphatic matter all along their course. On the other side of the Channel, around Boulogne, they have yielded a considerable quantity of low-percentaged phosphate of lime to commerce.

The production of the three counties of Cambridge, Bedford, and Suffolk, for the seven years named, is estimated as follows in the *Mineral Statistics of the United Kingdom*, edited by Mr. Robert Hunt, F.R.S.

	Tons.	Value.
1875	250,000	£627,000
1876	258,000	625,000
1877	69,000	200,000
1878	54,000	150,000
1879	34,000	73,750
1880	30,000	70,950
1881	31,500	86,628

PHOSPHATIC DEPOSITS OF THE ARDENNES AND THE MEUSE, FRANCE AND BELGIUM.[1]

Phosphates have been worked regularly in the Ardennes since their discovery in 1852 to 1855 by Messrs. de Molon, Thurneisen, Rosseau, and Dessailly.

The discovery soon passed from the limits of the Ardennes into the department of the Meuse, where 1,000 tons were raised in 1862, 11,000 tons in 1867, and at the present time the production is estimated at over 40,000 tons a year.

The strata of the region are divided in ascending order into the Greensand, the Gault, and the Tufaceous chalk or Gaize. A comparison of this description with the section fig. 34, show that these deposits correspond to those in the same strata in England. In the Ardennes the Greensand beds repose sometimes upon the Kimmeridge clay, and sometimes upon beds of older Jurassic age than this. In the arrondissement of Vouziers they rest upon the Kimmeridge clay, and they are developed in great strength in the communes of Grand Pré, Marcy, Cheviers, Sommerance, Fleville, Cernay, Chatel, Apremont, and Exermont. The formation extends in isolated patches to the north-east of these localities in the communes of Livry de Fosse, Remonville, Andevarmen, and Barricourt, where it is found resting upon the *Astarte* bed, a bed lower down in the series than the Kimmeridge clay. It is also found, but of less thickness, in the valley that descends to Quatre Champs and Vouziers, and in the communes of Ternon and Voncy it rests immediately upon the Astarte bed. It advances in very fine beds upon the coral rag of Saulces, Faissault, Puisseux, and Norron. To the west of this last commune it is reduced to nothing. It rests upon the Oxford clay and upon the inferior or lower Jurassic beds in the communes of Neuf, Maison, Aubigny, Logny, Bogny, and to the north upon the plateaux of Rumigny and the neighbouring communes. It will thus be seen that the Greensand beds rest in

[1] Ed. Nivot, *Notice sur les Gisements et l'Exploration des Phosphat de Chaux Fossiles dans le Département de la Meuse.* Bull. Ac. Royal, Belgium, second series, t. xxx.—xl., 25—40.

depressions of the Jurassic strata in which the latter have been more or less worn away.

In the Meuse the Greensand beds extend parallel to the two banks of the Oise, a tributary of the river Aisne, from Montblainville on the northern limit of the department to Clermont in Argonne. In all this region they repose upon the Portland limestone that forms the solid rock of the valley.

The total thickness of the Greensand in the Ardennes and the Meuse is from 90 to 140 feet. The superficial area covered by these beds is estimated at 36,800 hectares.

In the School of Mines at Paris there are samples from all the points worked within this area. The phosphate beds occupy two distinct levels, one in the Greensands and one in the overlying Gault. The first of these beds, or group of beds, is encased in ochrey and greenish sands which range from 9 to 16 yards in thickness. This whole deposit is formed of fine-grained sands and clays, which are mixed more or less with a great quantity of silicate of iron. The general composition is silica 52·00, protoxide of iron 28·00, with 7·00 to 8·00 of alumina, and the same proportion of magnesia.

There are two groups of phosphatic nodules contained in these beds. The first group is composed of white and grey nodules of the shape of nipples. They range in size from that of a walnut to that of a fist. They are very compact and of a metallic lustre. The interior is formed of an agglomeration of little grains of green chlorite in a phosphatic cement, a certain number also appearing in the encasing rock. The other group consists of black or dark green nodules often cemented together and penetrated around the outside with grains of the enclosing rock, and with crystals of iron pyrites or of gypsum. They are also mixed with iron pyrites and are impressed with casts of shells and of serpulæ, with traces also of coraline forms. The structure of the nodules of this group is more compact than that of the first. They are richer also in phosphatic matter, containing 55 to 60 per cent., while those of the first group only contain 40 to 45. In this respect, therefore, it will be seen that they resemble the two groups of the phosphatic nodules of similar beds in Russia, as described on page 158, except that these two classes of nodules occur in the same bed. The thickness of the bed is usually from 6 to 8 inches, sometimes not more than from 2 to 4 inches. It differs a little in its exact place in the Greensand, being sometimes quite at the base and sometimes a little higher up in the series. The extraction is by open workings of a usual depth of 6 to 7 feet. At Jardinet, west of Varennes, the workings extend to a depth of 15 to 20 feet. The nodules receive from the workmen the not over polite name of 'dung of the devil and his rogues.'

The Gault lies upon the Greensand, and is about 50 feet in thickness. It is composed of green and greyish green clays, sometimes brown, dark brown, and nearly black. These clays are plastic, and are used for the manufacture of tiles and pottery. They are sandy at their base, and pass gradually upwards into clay; balls of iron pyrites are scattered throughout them. Phosphatic nodules are also disseminated through the mass. These consist of corals and sponges, and the phosphatic shells common in the lower bed. A great number of these shells are ammonites and hamites, with some gasteropoda and lamellibranchiate shells. There are also teeth of fish, and fossil wood. These nodules are as rich in phosphoric acid as are the dark-coloured nodules of the bed in the Greensand.

Above the Gault is the deposit locally known as La Gaize. This occupies the place of our Upper Greensand. It is here a lenticular deposit intercalated between the Gault and the chalk marls. It forms a chain of escarpments about 900 feet high, which follows the left bank of the Oise. It is a tender and porous rock of a whitish yellow colour, passing in its lower part to a grey rock much more clayey than the upper part.

The phosphatic nodules lie about 15 yards above the clay of the Gault. They are known as 'nodules de Gaize' or 'coquins de Gaize.' They range from a dark grey to black in colour, and have a polished surface. Their interior is formed of greyish matter, in appearance like the surrounding rock. The nodules are similar to those of the Gault just described, and the fossils are nearly identical.

The nodules of the Ardennes and the Meuse yield a maximum of 25 per cent. of phosphoric acid, equal to 55° of phosphate of lime. In general composition they are similar to those from our Bedford and Cambridge deposits. Probably they contain more iron and alumina, and from their inland position can hardly be exported successfully.

PHOSPHATIC DEPOSITS OF BELLEGARDE, LA PORTE DU RHONE.

These deposits of the south of France are of similar character to those just described. They were referred to by Brogniart in 1822, and by various writers up to Risler in 1872. They occur in the environs of Bellegarde, Lanorans and Mussel being the principal localities where works have been carried on. The deposit worked answers stratigraphically to the base of the Gault and the phosphatic matters contained in fossils and their fragments, some of which have a rolled appearance. There are few if any nodules of a coprolite appearance. The proportion of phosphatic fossils to the entire mass is about one-fifth, and experiments show the following results with regard to the fossils named.

	Phosphate of Lime.	Carbonate of Lime.
Ammonite	46·20	22·80
Inocerame	38·25	33·55
Gryphea	52·00	24·25
Nautili	65·30	29·60

Dr. Voelcker gives the following analysis of a sample of Bellegarde phosphate, and he states that the sample was lighter in colour than the Cambridge nodules and was softer, and so more easily ground to powder.

Moisture and water of combination . .	2·79
* Phosphoric acid	25·10
Lime	40·11
Oxide of iron and alumina Fluorine . . . . }	14·38
Insoluble silicious matter	17·62
	100·00

* Equal to 54·79 per cent. of phosphate of lime.

Deposits of similar age to those last described are also worked between Montpellier and Avignon. The order of the strata, in descending order, is:—

1. Diluvium.
2. Marls and sands, with occasional nodules of phosphate reaching to a thickness of 80 yards.
3. Sandy clay, Gault proper, from 7 to 8 yards thick, with green grains of glauconite. This is sub-divided into—
   (*a*) *Bank superior*, yellowish gray clay, about 2 ft. 9 in. thick.
   (*b*) *Middle*, bluish green clay, about 2 ft. thick, divided from the next below by 6 ft. 6 in. of green sandy marl.
   (*c*) *Inferior*, composed almost entirely of friable shells among a green clayey sand. Base of Gault.
4. Aphen, superior, 6 yards } Greensand.
5. „ inferior, 16 „ }
6. Phosphatic nodules.

In this bed the phosphatic matter is concentrated in the débris of fossils. The bed is largely made up of fossils, mostly in a rolled form; but many shells are found in a perfect state of preservation, and they assimilate in their general character to those of the Gault. There are few, if any, of the rolled, shapeless nodules, designated in England as coprolites. It is interesting to notice the proportion of phosphate and carbonate of lime contained in each of the principal shells.

The general colour is lighter than that of the Bedford and Cambridge deposits, and the quality is higher than that of those deposits, and also those of the Ardennes. Dr. Voelcker gives the two following analyses:—

	No. 1.	No. 2.
Moisture and water of combination	2·79	2·95
* Phosphoric acid	25·10	27·76
Lime	40·11	41·88
Oxide of iron and alumina Fluorine	14·38	10·56
† Carbonic acid		7·10
Insoluble silicious matter	17·62	9·75
	100·00	100·00
* Equal to tribasic phosphate of lime	54·79	60·60
† Equal to carbonate of lime	—	16·14

PHOSPHATIC DEPOSITS OF THE CRETACEOUS STRATA OF RUSSIA.

In the *Journal d'Agriculture Pratique* of the year 1872 M. Yermelow drew attention to the rich deposits of phosphate of lime existing in the cretaceous rocks of Russia.

The deposits extend between the rivers Desna and Don, and they traverse the governments of Smolensk, Orel, Koursk, and Voronife, presenting a level line of length of about 100 miles. Along this line a vast quantity of phosphate of lime is estimated as available for working.

The discovery of these Russian deposits of the mineral, as far as their application to agriculture is concerned, dates from the year 1858. In the earlier part of the present century geologists, among whom was the late Sir Roderick Murchison, had noticed in the neighbourhood of the towns of Koursk and Voronife blackish stones, which they took to be ironstones. These stones, known by the popular name of samorod—black-stone or hornstone—had been worked from time immemorial, for the construction and repairs of streets and roads.

In the year 1850 M. Kiprianow, a Russian engineer, who had used the stones for this purpose, gave an account of his observations in the *Gazette de Koursk*, in which he spoke of the mineral as iron. He at the same time sent to various learned men samples taken from the deposits.

In 1858, as the result of the analyses and researches of M. Khodnew, Professor of Chemistry at St. Petersburg, it was ascertained that the samorod was largely composed of phosphate of lime, and in different places of varying proportions of carbonate of lime, oxide of iron, and alumina, all of which were mixed with the clay and sand that constituted the rock.

In 1861 M. V. Solsky, in the *Revue Agricole*, contributed a series of articles on the agricultural value of the deposits, while MM. Claus and Guilemin had, by their researches and analyses, arrived at similar conclusions.

In 1866 Professor Engelhardt, of St. Petersburg, accompanied by M. Yermelow, received the official mission of exploring the deposits, and from their reports we gather that the deposits lie under the white chalk, and are continued down into the greensands and sandstone below, the general descending order being :—

1. Soil.

2. Diluvial beds.

3. Clayey marl of a varying thickness.

4. White chalk.

5. Sandy marl, with phosphatic nodules scattered throughout the mass.

6. Greensands, in which are one or more beds of phosphate of lime from 6 to 12 inches thick, in the form of nodules and concretions, which are often cemented together.

The number of beds ranges from one to seven ; but of the higher number there is seldom more than two of importance, the rest being simply strings. The phosphatic nodules of the beds are intermixed with grey, brown, and yellow sands. The depth of the beds from the surface is very variable. Along its outcrop the nodules are mixed with the surface soil, while at a distance of a few hundred yards along its dip they are a good depth ; but as the strata rise again to the surface, forming a shallow trough or series of basins, the maximum depth is not very great.

The general direction of the beds is from north-west to south-east, from Koursk to the little town of Koomy. In the north-west portion of this belt the chalk beds are much developed, so that it is difficult to reach the deposits in depth. To the south-east the beds of the greensand prevail, and the deposits are more accessible. The encasing rock, whether of chalk or sand, both above and below the deposits, contain phosphatic matter, those below containing the largest proportion, and are most compact in their character.

The character of the nodules is variable, each deposit or locality having its special features ; but these may be broadly divided into two very distinct groups.

The first presents the form of nodules, round or kidney-shaped, of variable size, black, brown, grey, and green in colour. To this series belong the separate nodules, which are usually less rich. The second is an agglomeration of very large nodules cemented together into a sort of flag, which used to be quarried for road purposes. These nodules are richest when most dense, and of a deep black colour, the sandy, friable varieties being comparatively poor. The density, texture, and colour vary in different portions of the same beds. The cement enclosing the nodules is also phosphatic—numbers of fossils and fragments of fossils, bones, shells, corals, sponges, and wood. These are taken as belonging to the age of greensand, and they are richest in phosphoric acid—30 to 35 per cent.

In the governments of Tambow and Spaask the principal phosphate bed is covered by a bed of greensand, with grains of glauconite and nests of mica. We have seen how closely associated that last mineral is with the apatite of the older rocks of Canada and Norway. The following list will show the variations of the proportions of phosphatic matter in different parts of the area described :—

	Phosphoric acid	equal to	Phosphate of lime.
Zorino	14·47		31·59
Korennaya	17·90		38·78
Yablonetz	22·07		47·81
Kotawetz	27·24		59·01
Koursk	14·25		30·08
Nendowistche	16·11		35·17
Orlinoye Guesdo	18·48		40·35

The following analysis by Dr. Voelcker shows the general composition of the merchantable qualities of these phosphates :—

Moisture and water of combination	3·55
Phosphoric acid	22·42
Lime	33·84
Oxide of iron, alumina, fluorine, carbonic acid, &c.	9·94
Insoluble silicious matter	30·25
	100·00

These deposits occupy the same geological horizon as those of Bedford and Cambridge.

CHAPTER X.

PHOSPHATE OF LIME—continued.

Phosphate of Lime in Tertiary Strata—Phosphatic Deposits of Nassau, North Germany—Situation—Geological Structure of the District—Illustrations of Modes of Occurrence—Whence derived—Analyses—Phosphates of Tarn-et-Garonne, France—Growth of the Industry—Geological Position—Modes of Occurrence—Similarity to the German Deposits—Analyses—Phosphatic Deposits of Carolina, America—History of the Discovery of—Geological Position—Characteristics—Land and River Phosphates—Analyses—Recent Phosphates—Alta Vela—Aruba Island—Navassa Island—Pedro Keys—Redonda Island—Sombrero Island—St. Martin's Island.

THE PHOSPHORITE DEPOSITS OF NASSAU, NORTH GERMANY.[1]

IN the year 1850 M. F. Sandberger had distinguished the phosphate of the neighbourhood of Diez as a mineral of manganese. Later M. Meyer, in searching for manganese in the neighbourhood of Staffel, discovered a stony mineral which eventually proved to be phosphate of lime. The working of mines at Staffel for this mineral dates from 1863, and samples were shown at the Paris Exhibition. In July, 1864, Professor Fresenius and M. Moh, of Coblentz, made analyses of the mineral, the results of which led to the discovery and exploration of new beds and deposits, and the whole question attracted the attention of eminent English agricultual chemists and gentlemen engaged in the manufacture of chemical manures.

In the summer of 1867 I was engaged in the examination

[1] D. C. Davies on 'The Deposits of Phosphate of Lime recently discovered in Nassau, North Germany.'—*Geological Magazine*, 1868, p. 262, *et seq.*

of about fifty mines and mineral properties containing phosphorite within the area to be described. The general results of this examination were published in the *Geological Magazine* of the following year.

To the description contained in that communication I may now add some of the more practical details and results which were then confined to a private report. The principal phosphorite deposits of Nassau occupy an irregular area, bounded on the north-east by the town of Weilburg, on the north-west by the Westerwald, on the east by the Taunus Mountains, and on the south by the town of Dietz. South of this point, as well as to the north-east of Weilburg, there are traces of the occurrence of the deposits; but from the nature of the underlying rock they are limited in extent. Inside of the eastern and southern boundaries of this district flows the river Lahn, which is made use of at various points along its course for the purpose of washing the mineral from the surrounding clay, as well as for the carriage of the washed material to the junction of this river with the Rhine at Oberlahnstein, about three miles above Coblentz. The section, fig. 36, will give a general idea of the geological structure of the district.

FIG. 36.—SECTION ILLUSTRATIVE OF THE GENERAL GEOLOGICAL STRUCTURE OF THE DUCHY OF NASSAU, GERMANY.

1, Porphyritic and Basaltic Rocks, generally crowned with a castle. 2, Shaly and slaty Beds (*Schieferstein*), much contorted and disturbed. 3, Red Sandstone Beds. 4, Limestone Dolomite. 5, Phosphate of Lime Phosphorite. 6, Clay (*Thon*).

The basement rock of the district (1) is porphyritic, varying in colour from dark to light grey and green; the green is thickly studded with cavities containing softer felspathic and calcareous matter, which, after long exposure to the atmosphere, disappears.

Upon this rock, in its many cavities, rests a thick suc-

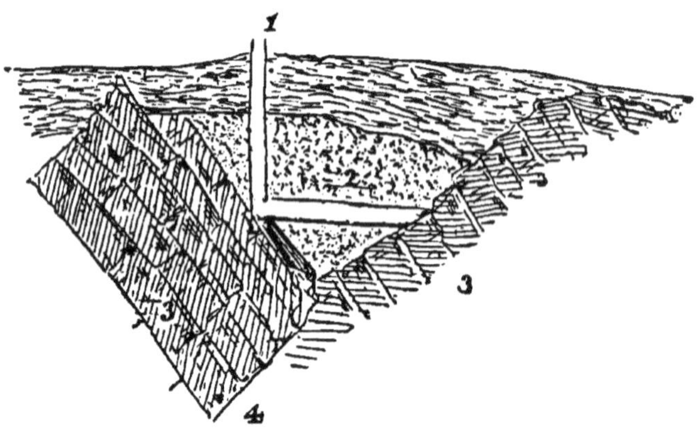

Fig. 37.—Section near Staffel showing Phosphorite resting in Dislocation of the Limestone.

1, Clay. 2, Phosphorite. 3 3, Limestone. 4, Line of Fault. 1, Shaft.

cession of slaty and shaly beds (2)—*Schieferstein.* These, as shown in an admirable section on the roadside south of Weilburg, are often greatly twisted and contorted. They are probably the equivalent of the slaty beds worked at Wissen-

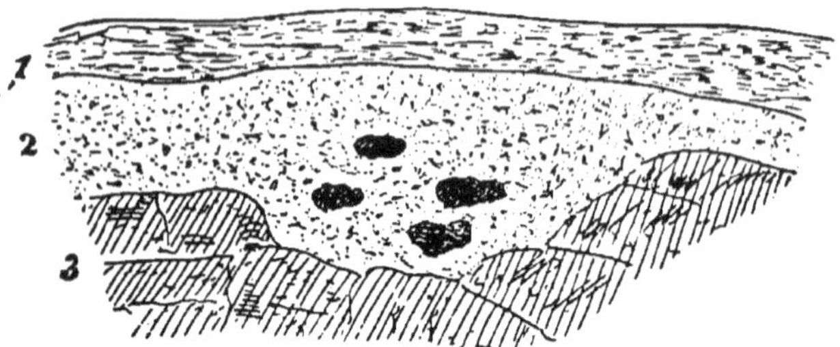

Fig. 38.—Section at Cubach, Nassau, showing Phosphorite and Manganese resting in Hollows of the Limestone.

1, Clay. 2, Phosphorite, 10 to 15 ft. thick. 3, Limestone. + + +, Deposits of Manganese.

bach to the north-west. These are overlaid by a great thickness of dark red sandstone beds (3) (*Spirifer Sandstein*, probably), which in places contains and is overlaid by hæmatite deposits, which are largely worked. Over large portions of the district these rocks are capped by a great thickness of massive limestone (4), locally known as *Dolomit*, and being probably the equivalent of the Eifel limestone and of the limestones in our middle Devonian series—the Ilfracombe slates and limestones. It is resting upon this, deposited in cracks and dislocations, fig. 37, and in water-worn hollows and abrasions, figs. 38 and 39, that the phosphatic deposit (5) is found, the whole series being crowned with a covering of brown clay (6) (*Thon*), which sometimes assumes a shaly appearance, and which also, in its upper portion, occasionally contains numerous fragments of the adjacent rocks.

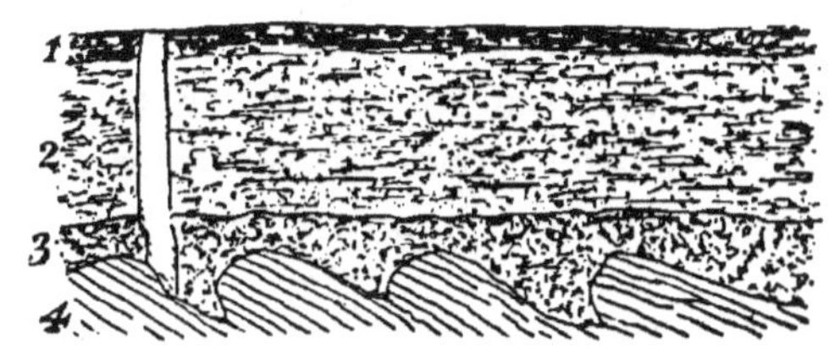

FIG. 39.—SECTION OF PHOSPHORITE NEAR ARFURT, NASSAU.

1, Clayey gravel. 2, Stiff Clay. 3, Phosphorite resting upon rounded and fretted edges of Limestone (4).

The deposit occurs in the form of concretions embedded in a matrix of clay. These concretions are most irregular in shape, and they vary in size from that of an apple to great masses, conglomerations of concretions, weighing several tons. It would also seem as if some of the original concretions had subsequently to their formation been subjected to a good deal of attrition. This is indicated by the preponderance of small fragments, decreasing in size to that of grains of sand. Where the deposit assumes this form it is known locally as *Washstein*.

Besides the phosphate of lime there are also deposits of hæmatite and manganese occurring with it in just the same position, and resting within the inequalities of the underlying limestone. As far as my observation went these deposits are found in bulk around the outer margin of the northern half of the phosphatic area, although there are some of considerable size in the more central portions of the district. It would also seem that some portions of these two minerals were held in suspension or solution by the water, and deposited along with phosphatic matter. They give the deep yellow and brown colours to some of the concretions, increase their hardness, and have so permeated the phosphatic matter in places as to considerably lessen its commercial value.

Along the north-western and north-eastern boundaries of the area, where the deposits border on the development of the older rocks, we find the greatest admixture of these extraneous matters, and the percentage of phosphate of lime ranging below 50 per cent.; but southward, on the great mass of limestone extending from south of Weilburg to Limburg, Staffel, and Dietz, the deposit improves in quality, is white and creamy in colour, and contains in places, as at Staffel, as much as 92 per cent. of phosphate of lime, when it yields some beautiful crystalline forms of apatite.

As might be expected from the mode of its occurrence, the

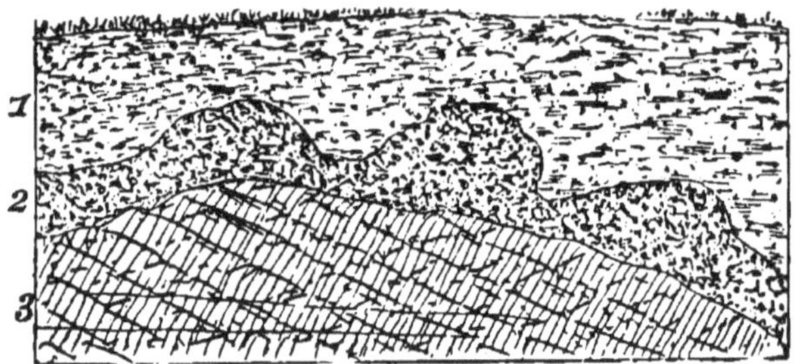

FIG. 40.—SECTION OF PHOSPHORITE WORKING IN OPHEIM, NEAR LIMBURG, NASSAU.
1, Clay, 30 to 40 ft. thick. 2, Batch of pale-coloured phosphatic Concretions, varying from 3 to 10 ft. in thickness, the Phosphate set in Clay and interspersed with a little Manganese. 3, Underlying Limestone.

deposit is very irregular in thickness, varying in the same mine from 6 inches to 10 feet. Fig. 40 represents its appearance in a mine at Opheim, a little to the west of Limburg. Generally speaking it attains its greatest thickness on a line ranging north-east and south-west along the centre of the underlying limestone, and it thins out gradually to the north-west and south-east. To the north-west and west the brown clay also becomes thinner, and is found covered with a splintery gravel (*Quartzgeschiebe*), the detritus of the neighbouring rocks. Subject to local variations, the phosphatic deposit seems co-extensive with the area of the limestone, its presence or absence at particular points depending upon—1st, whether there is a ridge or depression, or series of depressions, in the limestone; 2nd, the presence in force of hæmatite and manganese deposits; and, 3rd, the possibility of its having suffered denudation in exposed places since its deposition.

The clay is from 10 to 100 feet thick, and the method of mining is, when the clay is thin, to strip it off and work the deposit in an open work, as at Cubach, fig. 38; when the clay is thick a number of small shafts, about 4 feet diameter, are put down and communicated with each other by following the deposit underground. If the clay is wet or sandy, these shafts are secured by wickerwork. They are worked by windlasses.

The workings underground are most irregular, their direction being dependent upon the presence or otherwise of the phosphate. At only one group of mines, 'Heckolshausen,' did I see any attempt at artificial ventilation, and there the workings were low and wet, and the air very bad.

I have assumed at the head of this chapter that the deposits are of Tertiary age, and I would place them among the oldest of the Tertiary deposits—older than those of Tarn-et-Garonne, to which, in some respects, they bear great resemblance.

To the inquiry, Whence came such an amount of phosphatic matter? Several answers have been given. It has been supposed to have been derived from immense shoals of fish and other organisms which crowded the shallows of the limestone sea, and whose remains were deposited in the hollows and crevices of the rock. It has also been suggested that the phosphorite owes its origin to the emissions from below bringing up phosphoric acid; and, further, that the phosphate was dissolved out of the porphyritic rocks, as well as out of the limestones, by the action of carbonic acid. Primarily we must, I think, call to our aid the influx from below of phosphoric acid, aided, secondly, by the supplies derived from the dissolving of the older rocks. These two sources being compatible with the idea that life would be abundant in a sea of the geologic age, this would be charged largely with phosphatic matter, so that there was first of all an abundance of this essential element of organic life. The growth and decay of organic life would, in its turn, help to increase the quantity of phosphatic matter. There are not, it is true, any distinct organic remains in the deposits; but the structure and shape of these may have been obliterated by chemical action.

The specific gravity of the different colours is as follows:—

Pale buff.	Grey.	Dark, hard.	Dark.
1·9	2·6	2·7	2·8

the quality deteriorating with the density of colour.

The following are some of the mineral analyses:—

—	Weinbach.		Cubach and Edelsburg.	Staffel.		Graveneck.	Opheim.	Wilmar.
	Selected.	General.		Selected.	General.			
Phosphate of lime	58·40	54·00	53·30	92·00	62·00	57·60	65·10	52·70
„ of iron and alumina .	10·00	10·00	17·20	—	18·00	10·40	13·00	11·60
Sand and insoluble matter .	12·50	36·00 (Sand and insoluble matter, and sulphate of lime, together)	14·70	3·00	7·00	15·20	21·90	35·70
Sulphate of lime .	19·10		14·80	5·00	12·00	16·80	—	—
	100·00	100·00	100·00	100·00	99·00	100·00	100·00	100·00

The following are more recent and more detailed analyses of the richer sorts by Dr. Voelcker, of specimens from the neighbourhood of Staffel:—

—	No. 1.	No. 2.	No. 3.
Water	·65	·25	·98
*Phosphoric acid	40·56	38·12	36·19
Lime	56·29	53·92	49·44
Oxide of iron	1·21 (Oxide of iron and alumina together)	·93 (Oxide of iron and alumina together)	·96
Alumina			3·07
Magnesia	·97 (Magnesia and fluorine together)	·69	2·88 (Magnesia and fluorine together)
Fluorine		3·16	
†Carbonic acid	—	2·75	1·87
Sulphuric acid	—	·09	—
Silica	·32	·09	4·61
	100·00	100·00	100·00
* Equal to tribasic phosphate of lime .	88·54	83·21	79·01
† Equal to carbonate of lime	—	6·25	4·25

The first table, however, represents more correctly the average quality of the Nassau phosphorite; indeed, if we include the small stuff (washstein), the result is above the average. The cost of raising, dressing, and washing, with carriage to the river Lahn, amounts to 26*s.* to 30*s.* per English ton; the freight to Oberlahnstein 3*s.* to 4*s.* In order to send to England the mineral has to be reshipped on Rhine barges, and again reshipped upon seagoing craft. So that it will be seen that the German phosphates cannot be sent profitably to England, and, except with the higher qualities at the first, they have not. They are, however, largely manipulated at works established upon the Rhine.

THE PHOSPHATES OF THE DEPARTMENT OF TARN-ET-GARONNE, FRANCE.[1]

The important phosphatic deposits of these departments in the South of France attracted attention towards the close of the Franco-German war. In the year 1871 M. Daubre, of the Ecole des Mines, visited and described them, and since then an important industry has sprung up.

The phosphatic region is situated in the north-east of the department, on the right bank of the river Aveyron, in the neighbourhood of Montauban. The deposits occur on the summit of a great plateau which is interrupted by valleys of erosion. The basement rock of the country is Oolitic, principally the divisions Oxford clay and Coralline rag. From underneath these strata, at some distance, rise the granite and gneiss of Aveyron, and the district is not far from a recent volcanic region. Upon the Oolitic strata rest Tertiary beds, which are clayey and sandy beds of the Eocene strata.

Fig. 41 will shew the geological position of the phosphate bed with its contiguous strata.

The surface clay is yellow, red, and brown. It is strongly coloured by the oxides of iron, and intercalated in it are beds

[1] 'Phosphates de Tarn-et-Garonne,' par M. Lescure, *Bulletin de la Société Géologique de France*, Third Series, tom. iii.

of pisolitic iron ore, which have been much worked. Near the surface there are embedded in the clay the bones of living species of animals, and lower down there are numerous bones of extinct species, carnivorous and herbivorous, all huddled together, and strongly cemented as a breccia in a reddish clay. The clay on the surface does not give any indications of the phosphatic deposits below, but these are sought for where there are hollows indicating abrasions or fractures in the underlying rock.

The phosphate of lime is found in tubercular and kidney-

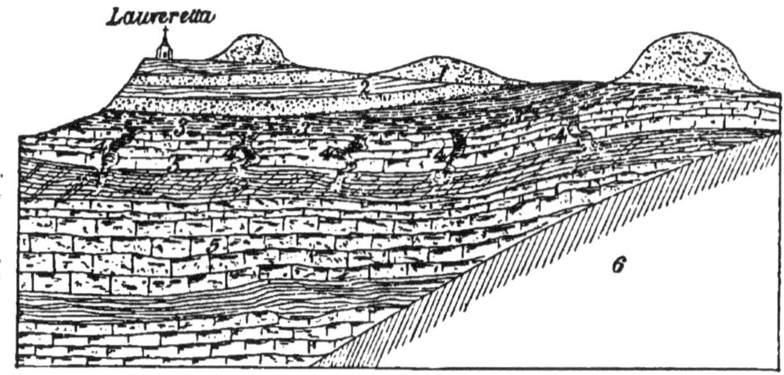

FIG. 41.—SECTION OF PHOSPHATIC AND ASSOCIATED STRATA, TARN-ET-GARONNE, FRANCE.

1, Lower Miocene Marls and Sands. 2, Eocene Marls and Sands. 3, Phosphatic Deposits resting on the Edges of Oolitic Strata (5). 4, Cracks and Fissures in Oolitic Strata charged with Concretions of Phosphate. 5, Coralline Limestone and Oxford Clay of the Oolitic Strata. 6, Liassic Strata.

shaped concretions in the cracks in the underlying limestones, as shown in fig. 41. They are intercalated vertically in the cracks in clay of reddish-brown and yellow colours, and often, also, occurring as thread or ribbon-like masses. They are sometimes white in colour, more frequently grey, with a waxy lustre and an opal-like appearance. Dr. Voelcker[1] says that he has found the white and opal-like specimens most rich in phosphoric acid, those of a yellow or brown colour less so—this being the

[1] 'On the Chemical Composition of Phosphatic Minerals used for Agricultural Purposes,' *Journal of the Royal Agricultural Society of England*, Second Series, vol. xi. part 2.

ordinary kind, while those of a dark brown colour are of an inferior kind. In these respects, as well as in the mode of their occurrence, it is interesting to compare these deposits with those of the valley of the Lahn, in Nassau, Germany.

The richest quarries or mines are those worked in the most vertical cracks, and those having a north-east and south-west direction. The quarry of Larnagol, in Lot, may be taken as an example of the rest of these mines. It is 35 kilometres north-north-east of Malperie, and is situated at a height of 360 metres on the summit of the plateau of Oxford clay. It is divided into three or four exploitations. It has a north-east and south-west direction; at the south-west end the clefts and veins running north-north-west and south-south-east are richest. From this end a large quantity of phosphate has been extracted. The veins at the north-east end are, or were recently, followed profitably.

The phosphatic concretions in these cracks have been supposed to owe their origin to geyserine ejections, also to infiltration of water charged with phosphatic matter derived from the bones in the overlying clay, and also to the same substance abounding in the lagunes of the Eocene sea. From the absence of organic remains, as well as of small particles of bony structure, we must, I think, attribute the origin of the deposit chiefly to causes linked with the first of these hypotheses, and regard these deposits as the result of phosphatic matter deposited pure and simple on the rocky floor of an Eocene sea from water largely impregnated with it. The lagunes would form deposits of a different character, resembling those at the summit of the London clay—that is, concretions around organic centres.

The phosphates from these deposits are known in England commercially as Bordeaux phosphates, being shipped from that port. At first the percentage of phosphate of lime ranged as high as 70 to 74 per cent.; subsequently it did not average more than 60 per cent.

The three following analyses, each of which I have selected

as the average of a considerable group given by Dr. Voelcker, will shew the composition of the mineral.

	High Quality.	Medium Quality.	Inferior Quality.
Moisture	3·01	1·64	6·81
Water of combination .	2·11	1·64	
*Phosphoric acid . .	34·01	30·47	25·96
Lime	46·77	44·69	31·51
Oxide of iron, alumina, carbonic acid, &c. . .	11·61	17·43	21·63
Insoluble silicious matter .	2·49	4·13	14·09
	100·00	100·00	100·00
*Equal to tribasic phosphate of lime . .	74·24	66·52	56·67

THE PHOSPHATIC DEPOSITS OF CAROLINA, AMERICA.[1]

These deposits, so important from their extent and commercial value, seem to have been first noticed by Ramsay in 1797, in the *History of South Carolina*, who spoke of them as remarkable discoveries of phosphate of lime. They again attracted attention in the year 1837. In November of that year, in a plain of rice about a mile from the river Ashley, in the parish of St. Andre, Mr. Holmes found a number of red nodules, very hard and covered with impressions of marine shells. These nodules were spread over the surface of the soil, and they were heaped up in places as so many stones, so as not to hinder the cultivation of the soil. Mr. Holmes having some knowledge of geology and palæontology, the shells, bones, teeth, and corals mixed up with the stones attracted his attention, and many of them were added to his collection of fossils. He pursued his studies and researches until the year 1842, when Mr. Ruffin was charged by the legislature to make an inquiry upon South Carolina from a geological and agricultural point of view, that gentleman having succeeded for some years in fertilizing poor land in Virginia with marl con-

[1] Brylinski, 'Rapport sur les Phosphates des Chaux de la Caroline du Sud,' *Société Géologique de Normandie*, tome ii., 1875.

taining about 25 per cent. of phosphate of lime. It was felt that the same results were possible in Carolina, and the farmers were eager for all the beds of marl and calcareous earth that could be found. Mr. Ruffin examined the country with great care and found extensive beds of marl. From samples collected from different localities he found marls containing carbonate of lime ranging from 50 to 80 per cent.

Nevertheless, the results obtained in Carolina from the application of these marls were not equal to those obtained in Virginia from poorer marls which were more easily attacked and dissolved by the liquid acids, while those of Carolina were so intimately mixed with silica, oxide of iron, phosphate of lime, and other substances. Upon calcining these Carolina marls, however, they were found to be more powerful fertilizers than those of Virginia. Among other samples the nodules found by Mr. Holmes were submitted to Mr. Ruffin, but not finding carbonate of lime in them to any extent, that gentleman regarded altogether as improper their employment as fertilizers.

About the same period, Professor Shepard, with Messrs. Lawrence Smith and W. Harmer, studied the question of the employment of marl in agriculture, but they also failed to distinguish the Ashley nodules from ordinary marl.

The experiments in marling went on, and searches for the material were made among other places on an old plantation near Charleston. In digging and proving the soil, at a depth of about two feet, a regular bed about one foot thick of rocky substances fixed in the clay was reached. These substances were covered with shells, and were evidently identical with the loose stones of the same character found spread over the plain.

Under this bed lay a marl bed of a yellowish colour, and containing 61 per cent. of carbonate of lime, passing into a marl of a greenish colour, containing up to 71 per cent. of carbonate of lime.

Mr. Holmes did not neglect the opportunity of studying and comparing the different rocks with those discovered by him, but without as yet arriving at any definite or useful results.

In 1848 a Mr. Tuomy paid some attention to the rock, and noticed especially the comparative absence of carbonate of lime in the lower marls, which he attributed to a different chemical condition of the water in which it was deposited.

In 1850, Mr. Holmes, in a paper read before the American Association for the advancement of Science, noticed the interesting character of the rock chiefly on account of the fossiliferous nature and of its fœtid odour; he also noticed the disparity between the amount of carbonate of lime in the rock, 2 per cent., and that of the underlying marl, 60 to 70 per cent.

Up to the year 1867, or thirty years after attention had been drawn to these deposits, the true character of the rock remained unknown, so that in 1859 Colonel Hatch amassed a great collection of the bones of extinct animals for the purpose of making manure, being ignorant of the properties of this nodular rock. So also in 1867 Messrs. Dakes & Co., who had formed with Dr. St. Julien and D. C. Ebaugh a company for the manufacture of manure, imported phosphatic rock from Nevassa, whilst they had close at home a large available supply of the mineral.

In 1867, Dr. St. Julien, having received from Dr. F. Glidding specimens of teeth and bones found in a plantation called 'The Elms,' belonging to Dr. Glidding's father, examined them carefully and recognised their true character. He therefore began to collect for himself the nodules found on the banks of the river Ashley. He sent one of these nodules to Dr. N. O. Pratt for analysis. This contained 34 per cent. of phosphate of lime. A specimen belonging to Mr. Holmes was found to contain 60 per cent. Mr. Holmes, foreseeing the value of these nodules as fertilisers, went to Ashley ferry, and studied the rock *in situ*, with a view of ascertaining the thickness and extent of the bed, and the results of his visit in these respects was eminently satisfactory. At this juncture the attention of Messrs. Pratt and Holmes had been directed by a book of the late Professor Ansted's to the phosphatic deposits of Cambridge, and they were struck with the resemblance between the character of the two deposits, as well as the apparent geological age, although in this they were mistaken. Convinced they had an important source of profit, they formed a company in Philadelphia, called the Charleston Mining and Manufacturing Company, with a capital of four million francs. The following table shows in descending order the geological position of these deposits, which is further illustrated by the sections, figs. 42 and 43.

	Age.
1. Cultivated soil. Man and living animals. Place of the glacial clays, sands, and gravels of Europe.	
2. Sands and fragments of shells . . . 3. Rocks of phosphate, yellow and greenish marl	Post-Pliocene.
4. Sands and clays containing fossil shells	Pliocene. The shells belong in a large proportion to existing species.

The lowest of these beds, No. 4, should in ordinary course rest upon Miocene strata, but these are absent, and it rests

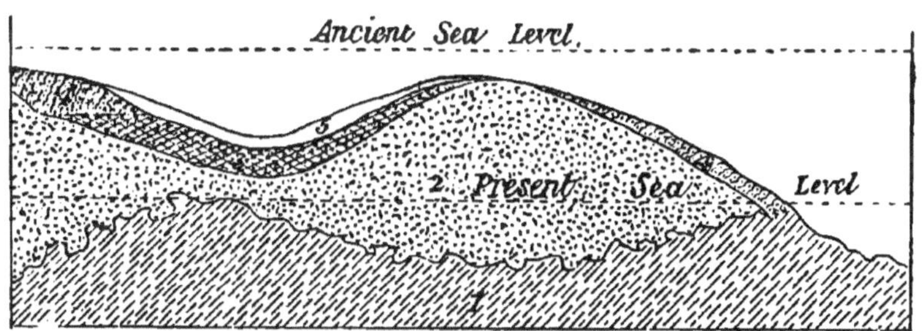

FIG. 42.—SECTION OF PHOSPHATIC DEPOSIT, SOUTH CAROLINA, SHOWING FORMER SEA LEVEL.

1, Eocene Marls. 2, Sand. 3, Phosphates. 4, Sand and Soil. 5, Salt.[1]

directly, as shewn in the sections, figs. 42 and 43, upon Eocene strata.

The organisms of the phosphatic bed consists of corals and coralline structures; the teeth and bones of marine animals resembling alligators. Few, if any, remains of terrestrial animals are found in the rock itself, although they are found among the nodules strewn upon the surface. These marine organisms seem to have been deposited in the deeper portions of an otherwise shallow sea, which probably was fed by the waters of the Atlantic bringing and leaving the molluscan, zoophytic, and crustacean life that abounded in its shallow waters.

The deposit is in hollows and pockets in the strata below, and varies in thickness from fifteen inches to three feet. The quantity of merchantable phosphate of lime contained in it is estimated at from 800 to 1,000 tons per acre. The deposit has been traced over an area of about 50 square miles, and it is not yet known how much greater its workable area may be.

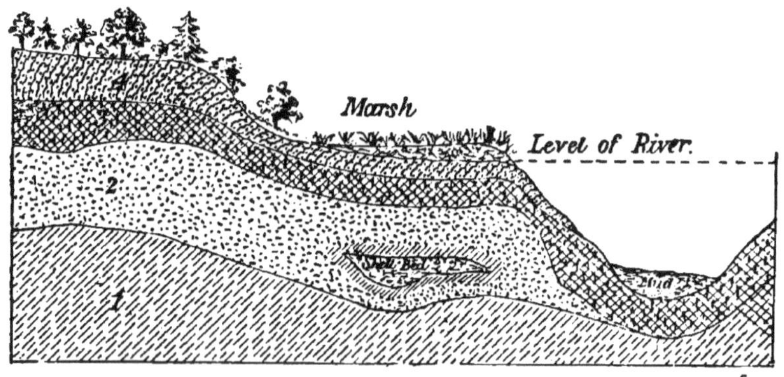

FIG. 43.—SECTION OF THE PHOSPHATIC DEPOSIT ON THE BANKS OF THE ASHLEY RIVER, SOUTH CAROLINA.

1, Eocene Marls. 2, Clayey Sand. 3, Phosphate Bed. 4, Surface Soil.

The bed just described is known as the land phosphate, but on the banks and in the beds of the rivers are the river phosphates of a similar character, and probably the same deposit as shewn in fig. 43. These also have been proved as extending over a large area, so that for extent and facilities for cheap working and transit these Carolina phosphates are unequalled by any other. The average quality, 53 to 54 per cent., is not high, but they are good workable phosphates. At the present time they form the chief source of our supply in this country, although the increasing demand for them in America is gradually effecting an increase in price which, if it progresses, will lead us to turn to the hitherto neglected sources of supply at home. The annual amount of Carolina phosphates imported into this country during the last ten years may be taken at 170,000 tons, of the value of £500,000.

Dr. Voelcker gives the following analysis illustrative of the composition of the Carolina land phosphates:—

	No. 1.	No. 2.	No. 3.	No. 4.	No. 5.	No. 6.	No. 7.
Moisture	7·40	2·29	10·30	3·98	8·01	6·59	7·69
Water of combination						1·09	1·34
* Phosphoric acid .	26·50	24·29	22·06	25·47	23·93	24·80	23·35
Lime	37·20	38·71	37·24	30·11	36·75	38·14	36·41
Oxide of iron, alumina, magnesia, carbonic acid, &c.	16·27	17·28	15·45	18·82	16·88	17·01	16·54
Insoluble silicious matter	12·63	17·43	14·95	11·62	14·43	11·67	14·67
	100·00	100·00	100·00	100·00	100·00	100·00	100·00
* Equal to tribasic phosphate of lime.	57·85	53·02	48·16	55·60	52·24	54·14	50·98

A full analysis by Dr. Charles H. Shepherd of Charleston, is also given below.

ANALYSIS OF GROUND LAND-ROCK PHOSPHATES MADE BY DR. CHARLES H. SHEPHERD, JUN., CHARLESTON, SOUTH CAROLINA, MARCH 22ND, 1879.

Moisture	0·48
Organic matter	4·30
* Carbonic acid	3·63
Sulphuric acid	2·18
† Phosphoric acid	27·24
Lime	42·68
Magnesia	0·57
Sesquioxide of iron	3·80
Alumina	1·10
Sand and insoluble silicious matter . .	13·17
Undetermined matter	0·88
	100·00
* Equivalent in carbonate of lime	8·18
† ,, ,, bone phosphate of lime . .	59·46

The reader interested in the working of English and Welsh phosphates may compare the above particulars, particularly the amounts of oxide of iron, alumina, magnesia, and carbonic acid, with the later analysis from bulk of the Welsh Silurian phosphorite given in Chapter VIII.

I have thus noticed the principal deposits of this mineral as they are found in the Laurentian, Silurian, Cretaceous, and Tertiary strata, and it only remains now for me to refer briefly to those apparently very recent deposits which are found on and surrounding the coralline islands of tropical seas.[1]

Alta Vela.—This is a small island near St. Domingo, from which phosphate of lime containing a good deal of alumina is obtained. Some samples do not show more than 43 to 44 per cent. of phosphate of lime. An intermediate sample tested by Dr. Voelcker gave the following result:—

Moisture	} 19·33
Water of combination	
* Phosphoric acid	26·33
Oxide of iron	7·23
Alumina	20·22
Insoluble siliceous matter	26·92
	100·00
* Equal to tribasic acid	57·26

Owing to the absence of lime and the presence in such large quantities of oxide of iron, alumina, and insoluble silicious matter, this phosphate, with that of Redonda, is not adapted to the manufacture of superphosphates. It is, however, used in chemical works for the production of alum, in the manufacture of which it yields an impure phosphoric acid which with salts of ammonia and other fertilisers may be worked up into chemical manures.

Aruba Island.—This is one of the Leeward Islands of the Caribbean Sea. It is situated in 12° 36′ N. latitude and 70° 8′

[1] Dr. Voelcker, 'On the Chemical Composition of Phosphatic Minerals used for Agricultural Purposes,' *Journal of the Royal Agricultural Society*, No. xxii., part ii. This paper contains very numerous analyses of different kinds of phosphates.

W. longitude. The island also contains gold, which has been worked more or less since the year 1824. The phosphate is a hard rock-like mineral from yellow to light brown in colour, with dark brown spots and bands, veins of calc spar. The quality ranges from 60 to 75 per cent., an average sample showing the following result:—

Moisture and water of combination . .	5·54
* Phosphoric acid	28·95
Lime	30·18
† Carbonic acid	·98
Oxide of iron	9·26
Alumina, &c.	17·22
Insoluble silicious matter	7·87
	100·00
* Equal to tribasic phosphate of lime . .	63·20
† Equal to carbonate of lime	2·23

Navassa Island is a small uninhabited island of the Caribbean Sea. It is situated in 18° 25′ N. latitude and 75° 5′ W. longitude. It is surrounded by coral reefs, the coral rock forming the framework, the cavities of which are filled with phosphatic matter of a reddish brown colour. This is made up of globular grains of phosphate of lime which are cemented together in hard masses. The phosphate varies in quality, but the following may be regarded as the composition of an average sample.

Moisture	8·50
Water of combination	4·15
* Phosphoric acid	28·47
Lime	34·07
Magnesia	·45
† Carbonic acid	2·30
Oxide of iron	4·49
Alumina	9·48
Sulphuric acid	1·81
Insoluble silicious matter	6·28
	100·00
* Equal to tribasic phosphate of lime .	62·15
† Equal to carbonate of lime . . .	5·22

Pedro Keys, a small island south of Jamaica, also yields the mineral, showing 60 to 65 per cent. of tribasic phosphate of lime, but mixed with 20 per cent. of oxide of iron, alumina, magnesia, and carbonic acid, the three first being the elements which lower the value of the mineral.

Redonda Island, situated 16° 54′ N., 62° 21′ W., is noted for the production of a phosphate of alumina which contains a good deal of oxide of iron and little or no lime, and can only be used for the same purposes as the Alta Vela phosphate already described. By a patent taken out by Messrs. Forbes and Price it was also intended to be used in the purification of town sewage.

An analysis by Dr. Voelcker shows the following composition :—

Moisture and water of combination	21·15
*Phosphoric acid	37·04
Alumina and oxide of iron	32·26
Insoluble silicious matter	9·55
	100·00
* Corresponding to tribasic phosphate of lime	80·86

Sombrero Island, one of the Leeward Islands, situated about 60 miles east of the Danish West Indian Islands, is one of the oldest sources of phosphate of lime in this region. The mineral is a light-coloured, nearly white substance, very light and porous, consisting of shells and other marine organisms of living species set in a coralline framework. The mineral has now to be obtained from below the water-line, and it must be difficult to work it profitably.

The following may be taken as its average composition:—

Moisture and water of combination	8·92
* Phosphoric acid	31·73
Lime	45·69
† Carbonic acid	5·99
Oxide of iron and alumina	7·07
Insoluble silicious matter	·60
	100·00
* Equal to tribasic phosphate of lime	69·27
† Equal to carbonate of lime	13·61

This is a very valuable mineral phosphate, and it is to be hoped that upon exploration other islands of the numerous groups to which those already enumerated belong, will be found to contain phosphate of similar quality.

St. Martin's Island is a small island belonging to the Windward Islands. It is situated in latitude 18° 5′ N., and in longitude 63° 4′ W. The phosphate from this island consists, like the Sombrero and Navassa, of phosphatic matter filling up a coralline framework. Its composition is variable, and great care has to be exercised in separating the richer quality from the usual minerals associated with it.

The percentage of phosphate of lime ranges from 37 to 81, of carbonate of lime from 6 to 47, and of oxide of iron and alumina from 1·14 to 11·97.

Doubtless the mineral will be found on many others of the neighbouring islands.

In reading the foregoing description of the phosphatic deposits of the world, it will be noticed how the earliest deposits in the Laurentian rocks of Canada and Norway are free from all traces of organic life, and how in all the succeeding deposits from the Silurian upwards, except in those of Estramadura and Nassau, organic remains are largely associated with the deposits and indeed form a part of them. This difference points to a difference in the original mode of deposition. In the first case the phosphatic matter was deposited pure and simple primarily on an ocean floor; and in the second case it had passed through organic forms and life before it found its place as a phosphate bed.

It is also interesting to notice that just as in the case of chloride of sodium a certain group of minerals clustered around it and became associated with it in its deposits, so with phosphorite another group—iron, titanium, manganese, and sulphur—are associated with it throughout all time, and how in its turn phosphorus is intimately blended, although in a less degree, with most of our iron ores—as it now seems—to their deterioration.

Practically, too, it will be seen that phosphate mining in

beds in the Secondary and Tertiary strata is a more certain and reliable operation, even where the mineral is only of ordinary strength, to apatite mining in the older rocks, where it occurs in strings, veins, and pockets, albeit of very high strength.

PART III.

CARBON, COMPOUNDS OF CARBON, AND SULPHUR.

(EACH MINERAL CONSTITUTING A CLASS BY ITSELF, FORMING CLASS IV. AND CLASS V. OF THE LIST IN THE CONCLUDING CHAPTER.)

CHAPTER XI.

CARBON AND CARBONACEOUS SUBSTANCES.

The Diamond—History of Attempts to consume it, by Boordt, Boyle, Cosmo III., Sir Isaac Newton, Sir George Mackenzie, and Sir Humphrey Davy—Diamonds of India, of Brazil, of South Africa—History of the Discovery and Progress of the Industry and Particulars of Mining—Notable Diamonds—Plumbago or Graphite of Borrowdale, of Ayrshire, of North Wales, of Ceylon—Particulars of Production in Ceylon—Graphite in America—Uses for which it is employed —Jet—Origin of Name—Jet Industry of Yorkshire.

THE DIAMOND, AND DIAMOND MINING.

THE diamond is pure carbon, and in this form is white and colourless. It is also tinged yellow, green, red, orange, brown, and black, when other minerals are present in it in minute quantities. It crystallises in several forms, some of them

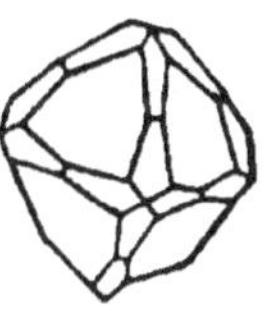
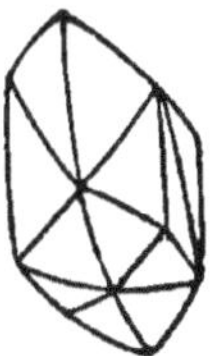

FIGS. 44, 45, 46, and 47.—USUAL FORMS OF DIAMONDS.

complex. Figs. 44, 45, 46, 47, show the ordinary shapes of diamonds.

It is transparent when pure, and has an adamantine lustre, but the darker kinds are opaque. H. 10, gravity=3·48 to 3·51.

Diamonds are known by their great hardness. Except in the instance I have given, where boron was obtained in a degree of hardness that would scratch a diamond, it is the hardest of all known substances. It was long considered to be incombustible, but in 1607 Boetius de Boordt suggested that it was inflammable. In 1673 Boyle proved that when it is exposed to a great heat it is dissipated into vapour. In 1694 Boyle's experiments were confirmed by those of Cosmo III., Grand Duke of Tuscany, with his celebrated burning-glass. About the same date Sir Isaac Newton, from its great refractive power, described it as an unctuous substance coagulated. Lavoisier proved it to be composed of carbon by throwing the sun's rays, concentrated by a powerful lens, upon a diamond enclosed in a vessel with oxygen gas. The diamond and the oxygen disappeared, and carbonic acid was generated. This experiment was repeated by Sir George Mackenzie in the year 1800. In 1814 the experiment of the Grand Duke Cosmo was repeated, with similar results, by Sir Humphrey Davy, in Florence.

Thus the hardest of all known bodies has been made to dissolve in the sun, and to pass away in a noxious vapour.

It is thought that the carbon of the diamond is of vegetable origin, having been dissolved and redeposited in some such way as those referred to in connection with the rarer redeposited forms of silica and alumina.

Diamonds were for a long time obtained almost exclusively from the East.

In India they have been found in the district of Golconda, near the Pass of Bezoara, on the northern bank of the Kistno, about fifty miles from the sea. The river passes through the hills by a narrow gorge, which in course of time it had cut for itself through the rocks. In doing this an extensive ancient lake beyond has been emptied, and it is in the drift once lying at the bottom of the lake, and derived from the wearing down of the neighbouring ancient rocks, that diamonds have been found. In digging through this drift there is first about eighteen inches of gravel, sand, and loam. Below this there is about four feet of mud and clay, and underneath this is the diamond deposit, three to four feet thick. It consists of a large number of rounded stones, and of gravel, cemented together in clay. Occasionally a thin layer of calcareous tufa occurs between this deposit and the overlying black mud. The natives were used to sink a pit a few feet diameter down to this deposit, and from the bottom to burrow in various directions as they best could. If unsuccessful, they speedily removed to another spot. From all accounts they seem to have made but a poor living at the work.

Another locality in India is on the left bank of the Mus-nuddy river, near its junction with the Mahanuddy. Here diamonds were found in a kind of ferruginous conglomerate, and this appears to be the nearest approach to finding diamonds in the solid rock. After all this deposit was a driftal one, like that to be described as occurring in Brazil, only it had been hardened by the presence of iron in its cementing matter.

Near Banaganpilly, 78° 4′ longitude, 15° 4′ latitude, diamonds have been found in a similar breccia, containing also hornstone, chalcedony, jasper, yellow and red quartz. This breccia passes into a conglomerate of pudding-stone, composed of round pieces of quartz cemented by a calcareous clay or mud. The strata of the nearest mountains consist of slaty rock of all kinds, with quartz rock, sandstones, flint and hornstone, pure limestone proper, and tufaceous limestone.

Diamonds are found in Borneo, on the west side of the Ratoos Mountains, associated with gold and platinum.

From the year 1728, when they were first discovered, until within the last twenty years, a prolific source of diamonds was the neighbourhood of Tejuco, on the district of Serro de Frio, or Cold Mountains, to the north of Rio Janeiro, in the province of Minas Geraes, Brazil.

The strata of the district consist of grits alternating with micaceous slate, with great masses of a kind of conglomerate or pudding-stone composed of grit and rounded quartz. In the lower lands along the river courses there is a finer conglomerate consisting of the same materials with fragments of hornblende and granite, all partly cemented together with oxide of iron. This is called *Cascalho*, and in it the diamonds occur, associated with garnets, topazes, and other precious stones, and gold. The streams are diverted so that the cascalho of the river bed may be excavated and washed, the water being removed out of the deeper parts by rude machinery, and in some places the cascalho is hoisted up by the same means. The deposit is then carefully washed by negroes in a series of troughs, the operation being also carefully watched by overseers who sit upon high seats that have no backs, in order to prevent the overseers from sleeping. Formerly when a negro found a diamond weighing 17½ carats he received his liberty, amid much ceremony. As much cascalho is excavated and stored during the dry season as is supposed to be sufficient to employ the negroes in washing during the months when rain is more plentiful. It is also found in practice that whether the diamonds be large or small, about the same weight or number of carats is found in each cubic yard of the cascalho. As a matter of course, there is a very large proportion of minute diamonds obtained. These are ground and used for polishing the larger ones. The average cost to the Government per carat of the diamonds found is said to be 33*s*. The principal deposits and workings are along the banks of the Jequitinhona River. From 1730 to 1814 the supply of diamonds from Brazil was estimated at 36,000 carats a year. In the year 1840 the yield was given at 20,000 carats. More recently the supply has been stated at from 25,000 to 30,000 carats yearly. At first it was difficult to sell diamonds obtained from Brazil, and the expedient had to be resorted to of sending them first to India, to be from thence sent to Europe as Indian diamonds.

Diamonds were discovered in the Ural Mountains of Russia, in the year 1829, by Humboldt and Rose, when travelling to Siberia.

SOUTH AFRICAN DIAMONDS.

The diamond fields of South Africa are situated mainly in Griqualand, about 700 miles north-east from Table Bay, and 450 miles inland from Port Elizabeth and Natal on the east coast. Kimberley is the centre of the principal mining field, and the chief mines near it are Kimberley, De Beers,

Du Toit's Pan, and Bultfontein. There are also two mines, Jagersfontein and Koffeyfontein, in the Orange Free State.

The discovery of diamonds was made in the year 1867 by Mr. O'Riley, a trader and hunter, while on a visit to a colonist named Van Niekirk.

The announcement of the discovery was at first received with great suspicion in England, one of the few men who believed in the genuineness of the stones from the first being the late Professor Tennant, of King's College, London. A writer in the *Geological Magazine* for 1868 concludes a rather long attempt to discredit the discovery with the words, 'I can only now conclude by expressing my conviction that the whole diamond discovery in South Africa is an imposture, a bubble scheme.' Nevertheless, the industry has grown to large proportions. Waterworks for the purpose of washing have been erected, and railways from the coast to the diamond district are in the course of construction. In the year 1881 diamonds to the value of 3,685,000*l.* passed through the post office.

The first diamond on being submitted to the authorities was valued at 500*l.* Considerable excitement followed the discovery, and the natives commenced searching for diamonds, in which they were successful. One diamond was found at 83½ carats, and was valued at 15,000*l.*

In the year 1868 the colonists searched for diamonds on the river Vaal, and succeeded in finding a considerable number. On the Transvaal side of the river the centre of the diggings is Klip Drift, and on the opposite side is Pniel. There are about fourteen river diggings in all. Du Toit's and Bultfontein mines were discovered in 1870. They are twenty-four miles from the diggings on the Vaal. De Beers and Kimberley were discovered in 1871, and in 1872 at the river diggings a great diamond of 182½ carats was found by Mr. Spalding.

The last four land mines lie within a reef which surrounds them on all sides. This reef is comprised near the surface of loose shale, which gives trouble to the miners by falling in. This is succeeded in depth by a trachytic breccia and augite. Below these are thin bedded strata to an unproved depth.

Inside the reef the surface soil is red and sandy. Below this comes a yellow calcareous gravelly deposit, under which the ground is of a blue and slaty nature, and in this lies the chief source of the diamonds. Diamonds of a large size have been found in the yellow ground, and the Kimberley mine has been productive from the surface, but usually these upper strata are not profitable to work.

At first in excavating the pits or holes, inclined roads, usually running north and south, were left up, but these gradually falling in, windlasses were resorted to, to be followed in 1873 by horse-whims. These gave way in the best mines in 1876 to steam-engines. In the early stages of working, when the diggers thought the red sand was the limit of the diamond deposit, the soil was simply turned over. When the underlying yellow ground proved to be productive the soil had to be removed a second time, and when the blue ground proved to be most productive of all, the same process had to be repeated at great cost. The mines or excavations are mostly irregular in shape, but some have been carefully laid out, and the shafts communicate with underground galleries. The Kimberley mine in 1882 covered 21 acres. The first diggers treated, on an average, eleven loads a day. Now, where a steam-engine is employed, a maximum of 250 loads a day is reached. The cost of working with present appliances is for the first hundred feet in depth 3*s*. 6*d*. per load, the second hundred feet, mostly in blue ground, 5*s*., the third 8*s*., and the fourth 11*s*. The yellow soil is loose and easy to work, but the blue requires blasting. Since mining operations commenced the price of labour has risen from 20*s*. per month to 30*s*. per week, with food. The following particulars, extracted from the published report of the De Beers Mining Co. of May 7, 1883, furnish much interesting information as to costs and results of working.

'DIAMONDS.—It is a matter of regret that the very severe fall in the diamond market during the last six months of the year has prevented your directors' anticipations as to the yield per load being realised, but the company's improved position will be apparent by comparison of this with last year's results, as it will be seen that during the year, to March 31, 1882, 96,439 loads ground washed yielded 76,859 carats diamonds, realising 104,552*l.* 8*s.* 8*d.*, whilst during the past year 166,436 loads washed yielded 149,396 carats, realising 158,675*l.* 4*s.* 3*d.*, showing nearly twice the output, and an improved quality of ground in the better average weight per load. The following is a summary of the work done during the year:—

Blue ground on floors, April 1, 1882 . .	3,000	16 cubic ft. loads.	
Do. deposited do. to March 31, 1883 .	179,785	"	"
	182,785	"	"
Blue ground washed, April 1, 1882, to March 31, 1883, 180,582 floors' loads, estimated at	166,136	"	"
(Discount of 8 per cent. being allowed for difference.)			
Leaving a balance on floors of	16,649	"	"

representing a cost, including rates, of about 5*s.* per load, or 4,000*l.*, which, however, with the cost of spreading lumps as given below, your directors have not considered right to include in the balance sheet, although a distinct asset of the company.

'The ground produced 149,396 carats diamonds, realising 158,675*l.* 4*s.* 3*d.*, giving, in spite of a fall of about 40 per cent. in the diamond market, an average yield of 19*s.* 1¼*d.* per load. The above weight includes 22,766 carats fine sand (17,032 carats found by the company, and 5,724 carats found on percentage). Besides the balance of blue ground as above, the company has 25,000 loads lumps spread out on its floors, representing a cost of 1,250*l.*, which are producing an average of two-fifths of a carat per load, showing that the ground, after allowing a percentage for black reef and high ground, has averaged for the year at least a carat per load.

'The cost of production, including rates, maintenance, and wear and tear of machinery, has been 11*s.* 9½*d.* per load, leaving a profit of 7*s.* 3¾*d.* per load. The following tables are given for your information:—

Blue Ground.			Floating Reef.			Washing.		Results.		Average Price.	
Gears.	No. of actual Working Days.	No. of Loads Hauled.	Gears.	No. of actual Working Days.	No. of Loads Hauled.	No. of actual Working Days.	No. of Loads Washed.	Diamonds in Carats.	Proceeds.	Average per Load.	Average per Carat.
No.			No.						£ s. d.		
1	137	42,225	1	143	49,539						
2	157	63,943	2	81	36,819						
3	187	55,482	3	108	36,654						
4	53	18,135	4	124	7,190						
			5	4	168						
4	312	179,785	5	287	130,370	305	166,136	149,396	158,675 4 3	19/1¼	21/3

Month.	No. of actual Working Days.	Diamonds Found, in Carats.	Proceeds.	Average per Carat.	Month.	No. of actual Working Days.	Diamonds Found, in Carats.	Proceeds.	Average per Carat.
1882.			£ s. d.	s. d.	1882.			£ s. d.	s. d.
April . . .	24	8,332	15,352 12 6	29 8	Oct . . .	26	14,676	15,252 6 0	21 1
May . . .	26	9,604	12,541 14 6	24 0	Nov. . .	25	14,718½	13,910 12 6	18 11
June . . .	26	8,751	12,075 0 0	27 6	Dec. . .	25	14,488	12,305 19 6	17 0
July . . .	24	10,286	12,488 8 6	24 3	Jan. 1883 .	26	15,459	12,999 5 9	16 9
August . .	26	13,217	16,574 19 0	25 0	Feb. . .	24	13,198	8,531 13 4	12 11
September .	26	12,152½	16,146 18 2	26 7	March . .	27	14,722¾	13,495 14 8	18 4
Total . . .	152	62,342½	85,179 12 8	27 4	Total . .	153	87,053¼	76,495 11 9	17 8

'SEARCHING SYSTEM.—The adoption of this system has caused a marked improvement in the company's finds. During the month of February, before the men were searched, the company's finds were, in the claims 658¼ carats, and on the floors 1,108¼ carats, whereas during the month of March, when the system had been put in force, they were, in the claims 1,234 carats, and on the floors 1,530¾ carats, making a total increase in the latter month of 998¼ carats.'

At the present time considerable interest is attached to the discovery of diamonds in New Zealand.

Among the great diamonds known may be mentioned—

1. One formerly belonging to the Crown jewels of France, weighing 67¾ carats.

2. The largest known, mentioned by Taverner as in the possession of the Grand Mogul, in form and size resembling half a hen's egg, and weighing 297 carats. This diamond was found in the mine of Colore, to the east of Golconda.

3. A beautiful lemon-coloured diamond, formerly belonging to the Grand Duke of Tuscany, now belonging to Austria, weighing 139 carats, and said to be worth 153,682*l.*

4. An eastern diamond, formerly belonging to Nadir Shah, Sultan of Persia. It is without flaws or defects of any kind. It is about the size and shape of a pigeon's egg. It was purchased by the Empress Catherine of Russia for about 90,000*l.* ready money, and an annuity of about 4,000*l.* in addition.

5. The Pitt or Regent diamond, said to have been found in Malacca. It was bought by Mr. Pitt, then Governor of Bencoolen, in Sumatra, and sold by him to the Regent Duke of Orleans, by whom it was placed among the crown jewels of France. It weighed 136$\frac{4}{16}$ carats.

6. The Koh-i-Noor, or Mountain of Light, belonging to Her Majesty the Queen, and formerly the property of Runjet Singh. This diamond is one of the largest in the world, and is valued at 2,000,000*l.* sterling.

The weight and value of diamonds are reckoned by carats, of which 150 are equal to 480 grains, or one ounce troy. The average price of rough diamonds is about 2*l.* per carat, but the price increases rapidly with size, as shown by the following table of approximate value.

					£
A rough diamond of		3	carats is worth		72
,,	,,	4	,,	,,	126
,,	,,	5	,,	,,	200
,,	,,	10	,,	,,	800
,,	,,	20	,,	,,	3,200
,,	,,	30	,,	,,	7,200
,,	,,	40	,,	,,	12,800
,,	,,	50	,,	,,	20,000
,,	,,	60	,,	,,	28,800
,,	,,	100	,,	,,	80,000

But the price will vary according to the demand, and at the present moment the diamond market is in a depressed state.

PLUMBAGO, OR GRAPHITE.

This mineral in its pure state is composed entirely of carbon, but it is usually found mixed with varying proportions of alumina, iron, lime, and other substances. Its uses are various. It is used in the manufacture of pencils, whence its name graphite. Mention is made of a document ruled with graphite in the year 1387. The mineral used for this purpose comes largely from Siberia, and Ceylon graphite was formerly used with that then obtained from the Borrowdale mines in Cumberland. For many years it has been extensively used in the manufacture of crucibles used in chemistry and metallurgy, on account of its great fire-resisting properties. More recently it has been applied to lubricating uses, and the commoner kinds for grate-polishing. The use of carbon for electric lighting purposes promises to open out a vast addition to the uses to which this mineral may be applied.

British Islands.—The most important graphite mine ever worked in England is that of Borrowdale, in Cumberland. It is recorded that 100,000*l.* worth of the mineral has been extracted in a year, the price then being 45*s.* per lb. The early history of the mine also shews the lawlessness of the times only a century

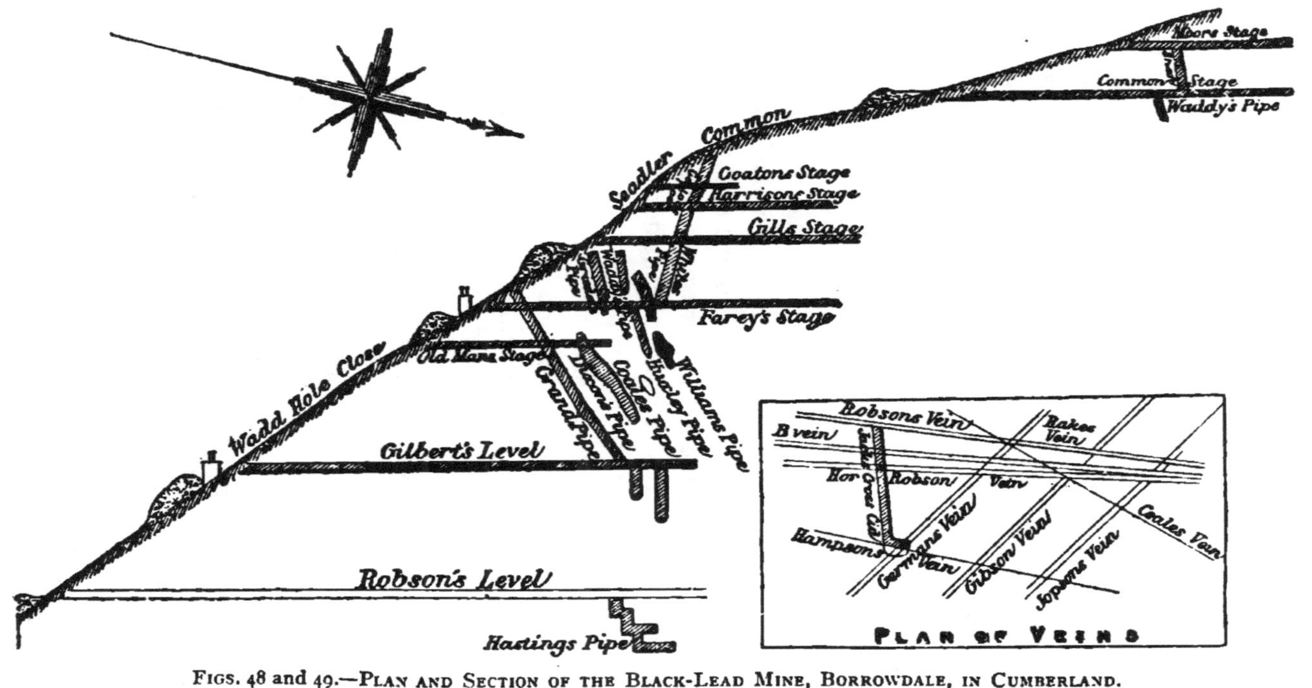

Figs. 48 and 49.—Plan and Section of the Black-Lead Mine, Borrowdale, in Cumberland.

Scale 60 fms. to 1 inch.

ago. A body of miners took forcible possession of the mine, overpowering the guard, which, in consequence of numerous robberies, had been placed to protect the property, the proceeds of which were sold to Jews, who came regularly to the George Hotel, Keswick, to buy. In the year 1800 the entrance to the mine was protected by a building containing offices and an attiring-room for the men. In this room was a trap-door through which alone an entrance to the workings was to be gained. Between the years 1850 and 1860 it was estimated that there was still about 30,000*l.* worth of the mineral in reserve, but this seems now to be exhausted. In the year 1875 20 tons were returned as being raised. The last return was for the following year, 1876, when one ton only was obtained. The mine was worked about half-way up a mountain 2,000 feet high. The strata was of Lower or Cambro-Silurian age. They consisted of slate rocks, with interbedded greenstone, and felspathic trappean rocks. The graphite occurred in one of these trap rocks of a bluish colour, in the form of irregular masses and more regular beds, which were sometimes of considerable size. These were connected by numerous veins glazed with plumbago. Both the felspathic bed and its enclosed graphite were frequently cut off or interrupted by rents, fissures, and also dykes of other substances, which rendered the mining of the mineral rather uncertain. A plan of the veins and a section of the workings is given in figs. 48, 49. It is on record that the mineral was first discovered by a tree being uprooted by the wind between Gills and Fareys stages. At first the plumbago was only used by the neighbouring farmers to mark their sheep. Afterwards its true value was discovered. In 1778 the price realised was 30*s.* per lb. In 1829 it was 35*s.*, and in 1833 45*s.* per lb. Afterwards, through the introduction of the mineral from abroad, probably Ceylon, the price fell to 30*s.* per lb. Some of the pipes and pockets of ore have contained as much as 30,000 lbs. of ore.

Early in the present century a graphite mine was worked near Cummock, in Ayrshire. The strata at this mine consisted of the following, in descending order:—

1. Sandstone composed principally of concretions of greyish white quartz, with a few scales of mica interspersed.

2. A bed of clay slate from 10 to 12 feet thick, passing in places into a flinty slate or basaltic hornstone.

3. Greenstone disposed in globular distinct concretions, which contained imbedded portions of graphite.

4. A bed of clay slate 12 feet thick, similar to No. 2.

5. Another bed of greenstone from 3 to 10 inches thick.

6. Graphite. This bed was from 3 to 6 feet thick, and was comprised of graphite and columnar glance coal. The graphite was found compact, scaly, and columnar. The glance coal was disposed in distinct columnar concretions, arranged like pillars of basalt. Both substances were intermixed, the graphite being included in the coal and the coal in the graphite; and in different parts of the patches of greenstone were met with what seemed to be part of the original deposit.

7. A layer of greenstone.

8. Flinty slate 10 to 14 feet thick.

9. Sandstone of similar structure to No. 1.

The graphite of Cummoch was more variable in quality than that of Borrowdale, but at the time it was considered that the deposit was extensive enough and the average quality good enough to warrant extensive workings. A graphite or black-lead mine, as it was called, was also discovered by accident, in the year 1816, in Glenstrathfarra, the property of Fraser ot Lovatt, who worked it for a short time. The rock in which the graphite occurred was gneiss of a micaceous character. It occurred in irregular masses, one of which was about three feet thick and several yards long. This was not throughout pure graphite, but was mixed with fragments of the constituents of the rocks—felspar, mica, quartz, with precious garnet. Numerous other masses of a smaller size were also found. The graphite was scaly, foliated, and undulatory. It also varied in quality. The working was carried on for a short time, and 5 tons were sold in the London market at 93*l.* per ton, the expense of getting it being stated at 13*l.* a ton. The work, however, came to an end, and at the present time there is not a graphite mine in the United Kingdom.

In North Wales, however, at a little distance above the Bala limestone, see fig. 26, there is a bed of impure graphite of considerable thickness. This bed follows the course of that limestone throughout Montgomeryshire and Merionethshire, see fig. 25. I have seen it near Llansaintfraid and Penygarnedd in the former county, and near Llanymawddy in the latter. It is worth the trial, as the bed is considerable, whether by cheap mining and careful washing this deposit could not be utilised for some of the purposes for which the mineral is in demand. A similar bed in like position occurs at Llangelenin, near Conway.

The chief source whence the bulk of the mineral has been derived for many years is the Island of Ceylon. The earliest notice there is of the mining of the mineral in the island is contained in a report by Colonel Colebrooke in the year 1829, where he says, relative to a tax, 'Provision had been made for payment either in money or in grain, and also for the delivery of cinnamon and *black* lead.' At that period the graphite of Ceylon was growing in repute, for we find that in America the late Mr. Joseph Dixon, who was the founder of the graphite industry and manufacture in the United States, started his manufacture in October, 1827, using a compact graphite found in New Hampshire; but seeing some of the specimens of the foliated variety brought from India by trading ships, he tested them, and procured a shipment, following it by another, and he finally adopted the Ceylon graphite entirely. The successors of Mr. Dixon, the Dixon Crucible Co., New Jersey City, New York, and Battersea Crucible Co., London, have hitherto taken by far the larger part of the production of the island. No record of the quantity of the mineral annually raised was kept before the year 1846. Since that date the quantity exported has been as follows:—

	Cwts.
For the five years ending 1851	13,410
" " " 1856	13,950
" " " 1861	37,530

		Cwts.
For the five years ending	1866	57,295
,, ,, ,,	1871	124,714
,, ,, ,,	1876	137,474
,, three years ,,	1879	114,671

It will thus be seen that the industry has been a growing one. The largest quantity exported in any single year was during the twelve months ending September 30, 1879, the quantity being 200,000 cwts. The quantity shipped in the year 1880 was nearly equal. The number of graphite mines and pits on the island is estimated at 400. The natives also often find lumps in the soil. The only information we seem to possess as to the geological conditions under which the mineral is found is contained in Dr. Gygax's *Geological Survey*, 1848, Appendix No. 2 to 'The Reports exhibiting the Past and Present State of Her Majesty's Colonial Possessions,' where the Doctor says, 'Plumbago or graphite is found chiefly in the southern side of Saffragam, in the Kukuls Korle. It is believed to belong to the same formation as the anthracite, viz., to the upper strata of the Devonian formation. The principal mine is at Nambepane, and contains a large vein running from north-west to south-east. The ore is pure and crystalline near the basalt, and compact and massive further from it. I believe this vein extends to a distance of forty or fifty miles towards the Bintenne country. The plumbago of Ceylon is pure and light, and now that a method has been discovered to purify and compress it the value will rise.'

As a matter of fact the price has fluctuated considerably. In 1868 and 1869 it fetched in Ceylon 12*l.* to 14*l.* per ton. In 1870, we are told in the provincial reports that the fall in the price is so considerable that it has put an end to the digging for the mineral on Government land. In 1872 there was a slight rise, and we read that in the Government of Galle the quantity raised was 22,751 cwts., and the average price 6 rupees per cwt., or 120 rupees per ton. From another province, in 1873, we are informed that 'plumbago, which formerly sold at 200 rupees per ton, is now only 90 rupees, and with the enhanced value of labour it can scarcely be profitably worked.' In 1874 the trade was at its worst, and we are told that plumbago 'is practically unsaleable.' In 1880 the average price was slightly higher than in 1868-69, being 15*l.* per ton. The deposits are spread over large areas in the Government of Galle, in the Hambantota district, at the Rannialakand Mountain, and the hilly country forming the north-western boundary adjoining Matara. There are also numerous mines in the Matara district, and also in the Wenda Willi Hatpatta. The Government grant licences to work the mineral at a royalty which was formerly 10 rupees per ton, but upon representations being made that this amount pressed hardly upon the poorer kinds of mineral, the royalty was reduced to 5 rupees. Shafts have been sunk, and attempts have been made to work the mines English fashion, but for the most part the deposits are worked open and near to the surface. The favourite mining district at present is the neighbourhood of Kurunegela, Awisawella, Ratnapura, and Kalutara. The natives are guided by lumps and grains in the soil, and by the croppings up of the rock. No geological survey has been made.

From 4,000 to 5,000 men are employed at the mines, and about 500 carters, with their carts and a pair of bullocks each, cart the mineral to Colombo. A good deal of the preparation of the mineral for the market is done at Colombo, women being largely employed. At first the Cingalese women had a strong prejudice against touching the mineral, but now they like the work, and are experts. One proprietor in Colombo, Mr. W. A. Fernando, whose family have long been connected with the trade, employs about 150 men and women. The men are paid from 50 to 75 cents a day, and the women 25 to 30. Mr. Fernando has large sheds, roofed with cocoanut-leaves—the dust blown about makes everything so slippery that slates would fall off. The plumbago is first washed in large baskets, the smaller pieces and dust being spread upon an asphalte floor to dry. By this means the quality is discovered by the practised eyes of the pickers, who separate pieces affected by iron ore, pyrites, quartz, or other foreign materials, a very small quantity of which would spoil the mineral for crucible-making. The good lumps have the dust brushed off, and are polished with cocoanut husks. The mineral is separated into four sizes by means of perforated plates of iron. It takes about 100 expert men and women to prepare about 3 tons a day of the smaller stuff. The ore is brought from the mines to Colombo in casks holding about 5 cwt. each, and also for shipment. Some 35,000 casks were required for this latter purpose in the year 1879.

UNITED STATES OF AMERICA.—The Dixon Crucible Com-

FIG. 50.—SECTION OF THE BLACKLEAD MOUNTAIN, TICONDEROGA, ESSEX COUNTY, NEW YORK.

1 1 1, Beds of Plumbago. 2 2 2, Gneissic Rocks.

pany, already referred to as large consumers of Ceylon graphite, have within the last six or seven years obtained the mineral also from a mountain locally known as the Blacklead Mountain, which rises close to the village of Ticonderoga, Essex County, State of New York. The graphite beds are interstratified between gneissic rocks, as shown in fig. 50. The beds dip at an angle of 45 degrees. The ore in them is chiefly of the foliated variety, and is mixed with gneiss and quartz in the beds, in veins or layers from 1 to 8 inches in thickness, some of the deposits being richer than others. One of these, as shown in the figure, has been followed to a depth of 350 feet. It is found of varying thickness, and it opens out at times into pockets. When separated from the attached materials this graphite is of fine quality. It is sent downhill from the mine to the works—a distance of two miles—where it is crushed with a stamp-battery, and the ore is then washed and separated in Cornish buddles and settling-tanks. The separated graphite scales are then ground in water to the fineness required for the different uses, as crucibles, lubricants, pencils, and ordinary blacklead. The industry is an old one. In the year 1822 the mineral was removed from the gangue by means of chisels, pickaxes, and iron bars, and conveyed to the falls, where it was pulverised and purified. In the manufacture of pencils the very finest grained graphite is used, and is mixed with clay. Graphite is found in various other localities in America—in the older rocks of the Appalachian Mountains, stretching from Canada to Alabama.

It is said to occur in great purity in five different localities in Albany county, Wyoming Territory, in veins from 1 foot 6 inches to 5 feet thick. At Pilkin, in Gunnison County, it occurs massive in beds 2 feet thick, but of impure quality. Indeed, it would seem that the massive beds everywhere were the most impure, the redeposited mineral in cracks and cavities being the purest. In New Mexico it is found in a pure form in the Coal-measures, possibly as the result of heat, which has driven all bituminous matter away, and of a redisposition of the particles. It is also found in the Black Hills of Dakota, and it has been mined at the Sonora mine, Tuolomme County, California.

It is also found in Canada, and the following table of analyses is from the report of a survey and inquiry authorised in 1876 by the Canadian Government as to the comparative merits of graphite from Canada and Ceylon. Graphite also occurs on the American side of Behring's Straits, where the natives use it for the ornamentation of their persons.

ANALYSES OF CANADIAN AND CEYLON GRAPHITES.

Locality.	Specific Gravity.	Volatile matter.	Carbon.	Ash.
		Per cent.	Per cent.	Per cent.
Canada, Buckingham, vein graphite, variety foliated	2·2689	0·178	99·675	0·147
Canada, Buckingham, vein graphite, variety columnar	2·2679	0·594	97·626	1·780
Canada, Grenville, vein graphite, variety foliated	2·2714	0·109	99·815	0·070
Canada, Grenville, vein graphite, variety columnar	2·2659	0·108	99·757	0·135
Ceylon, vein graphite, variety columnar .	2·2671	0·158	99·792	0·050
Ceylon, vein graphite, variety foliated .	2·2664	0·108	99·679	0·213
Ceylon, vein graphite, variety columnar .	2·2546	0·900	98·817	0·283
Ceylon, vein graphite, variety foliated .	2·2484	0·301	99·284	0·415

In Canada, New York, and Pennsylvania it occurs chiefly in veins and pockets in strata associated with gneissic rocks. It is associated in the veins with calcite, quartz, pyroxene, mica, and apatite; and the crystals of calcite, on being split, show scales of foliated graphite along the lines of cleavage.

The new report on the mineral resources of the United States, edited by Albert Williams, and published by the Government, gives the following particulars concerning the uses for which the mineral is employed, and it is stated that no less than 150,000,000 pencils are now manufactured in the world. The quantity of graphite imported into America in 1882 was 16,047,100 lbs., of which the greater part came from Ceylon, and the rest from Germany.

The properties of graphite make it useful for the following general purposes: the manufacture of refractory articles, lubricants, electrical supplies, pigments, and pencil-leads. A detailed table of the articles made from it is annexed, with an estimate of the percentage used for each purpose:—

PROPORTIONATE AMOUNTS OF GRAPHITE USED FOR DIFFERENT PURPOSES.

Manufactures.	Kinds of graphite used.	Per cent.
Crucible and refractory articles, as stoppers and nozzles, crucibles, &c.	Ceylon, American	35
Stove-polish	Ceylon, American, German	32
Lubricating graphite	American, Ceylon	10
Foundry facings, &c.	Ceylon, American, German	8
Graphite greases	American	6
Pencil-leads	American and German	3
Graphite packing	Ceylon, American	3
Polishing shot and powder	Ceylon, American	2
Paint	American	$\frac{1}{2}$
Electrotyping	American, Ceylon	$\frac{1}{4}$
Miscellaneous—piano-action, photographers', gilders', and hatters' use, electrical supplies, etc.	. . . .	$\frac{1}{4}$
		100

Graphite is also, as before stated, largely and increasingly used as a lubricant, both by itself and mixed with fat or grease, in various proportions.

JET.

In appearance this mineral resembles cannel coal; but it is harder, and is capable of receiving a high polish. It is also of a deeper black colour. It receives its name from the river Gages, in Lycia, in the alluvium of the mouth of which it was found in the time of Pliny. The small pieces of the mineral found there were called *gagates*, subsequently *gagat*, and ultimately jet. It has formed the subject of a considerable industry along the eastern side and seacoast of Yorkshire from very early times. The Danes, and subsequently the Romans, seemed to have worked it for ornamental purposes. A good specimen from Whitby, on analysis, showed the following composition:—

Carbon	79·97
Hydrogen	4·30
Nitrogen	0·47
Oxygen	13·22
Sulphur	0·91
Ashes	1·33
	100·00

It is found in the harder portions of the alum shale rock of the Lias. The rock is from 6 to 7 yards thick, and in this the mineral is found in small patches, and in quantities from a few ounces to two hundredweight. On the south it is found a little way south of Whitby, northward in the Mulgrave mines, then in those of Lord Dundas, terminating at Skinninggrove Beck. It then turns westward by Eston and the Guisbro' mines, and the mineral is found in all the valleys of the tributary streams of the Esk.

A softer kind is also found near Whitby, called soft jet, from which an inferior kind of article is made.

The mineral is obtained by excavating the face of the cliffs, and following it in old quarry workings. It is said that all attempts to mine systematically have been unsuccessful, although the price of the best qualities is from 12*s.* to 14*s.* per lb. It is found in Russia in sand and gravel beds, where it is called black amber, from its being electrical when rubbed, like amber. The production of jet in England in the year 1880 was 6,720 lbs.

CHAPTER XII.

CARBON—continued.

Asphaltum, Varieties of—History of the Uses of Bituminous Substances—Of the American Petroleum Industry—The Cannels of Flintshire and Lancashire—Modes of Occurrence—The Torbane Hill Mineral—The Bituminous Deposit of Bovey Tracey—Deposits of Bituminous Matter in Silurian Strata in Ireland—Bitumen in France—Gneiss in Sweden—Bitumen Deposits of France—Asphalte of the South of France—Bituminous and Petroleum Deposits of Germany—The Hanover Oil-well Region.

BITUMEN AND BITUMINOUS SUBSTANCES.

BITUMEN occurs in nature in various degrees of fluidity and solidity. The solid varieties are of a black, brown, or reddish brown colour. The fluid varieties are transparent, and vary from colourless to yellowish white and dark brown. The following are the chief varieties of bituminous substances.

ASPHALTUM (*Mineral Pitch*).—Chemical composition: 76 to 88 carbon, 2 to 10 oxygen, 6 to 10 hydrogen, and 1 to 3 nitrogen. Specific gravity, 1·0 to 1·2. Hardness 2. In colour and appearance pitch black, opaque, resinous. Has a strong bituminous odour.

MINERAL CAOUTCHOUC (*Elaterite* or *Elastic Bitumen*).—Chemical composition: 84 to 86 carbon, 12 to 14 hydrogen, and a little oxygen. Resinous colour, yellowish brown, and black. Occurs in kidney-shaped lumps; has a strong bituminous odour.

RETINITE (*Retin-Asphaltum*).—Composed of carbon, hydrogen, and oxygen in uncertain amounts; composition sometimes given as vegetable resin 55, bitumen 41. Occurs in roundish masses varying in colour, light yellow, brown, and green. Flexible and elastic when first obtained, but losing these qualities on exposure to the atmosphere.

PETROLEUM (*Fluid Bitumen*).—In its natural state it is dark yellow, brown, and dark brown in colour. Its chemical composition will be given in detail further on.

NAPHTHA or *Mineral Oil.*—Chemical composition: carbon 84 to 88, hydrogen 12 to 16; colourless to yellow. Before describing the conditions under which bituminous substances are found it may be well to give a brief general history of their use, more especially their use for lighting and heating purposes.

Perhaps the earliest reference we have is contained in the Scripture account of the building of the Ark and of the Tower of Babel. In the region assigned to that building bituminous matter is still abundant, and floats on the waters of the rivers. Herodotus, writing 440 years before the commencement of the Christian era, describes a place, Aderrica, situated thirty-five miles from Susa, where there were wells yielding salt, bitumen, and oil. The oil was drawn from these wells by means of skins, and was placed to settle in tanks. The more solid matter, salt and bitumen, fell to the bottom, and the oil was drawn off for use. The oil had an unpleasant smell, it was black in colour, and it was called by the natives Rhadinace.

Along the shores of the Caspian, and southward down the valleys of the Euphrates and Tigris, similar oils are still obtained and used by the inhabitants.

The oil from the wells of Rangoon, in Burmah, has also been in use from very ancient times.

About one hundred and thirty years ago the *Philosophical Transactions*, the *Transactions of the Royal Society of Great Britain*, and the scientific papers of other countries of Europe, contained references to the distillation of oils from coals, and of the experiments made towards purifying them. A century ago the Earl of Dundonald distilled these oils from coal in ovens similar to those now in use for the manufacture of coke. On the continent of Europe, a quarter of a century later, oils distilled from the tars obtained from bituminous schists by

Laurent, Reichenbach, and others, and purified to some extent by Selligue, were extensively used for burning and lubricating purposes.

In connection with the discovery and utilisation of gas, about the same time and subsequently, experiments were made and improvements introduced in the purification of the various oils obtained from coals.

In 1830 M. Selligue obtained a light paraffin oil from the bituminous shales of Feymoreau, in the Bourbon Vendée, France, together with a heavier oil and a lubricating oil and solid paraffin, by a method which is said to have been identical with that patented by Mr. James Young for the Torbane Hill mineral in the year 1850.

In America Dr. Abraham Gesner first successfully obtained lamp-oils from coals in the year 1845. These oils were burnt by him in lamps used by him at the public lectures given by him at Prince Edward's Island in the following year, and at others afterwards delivered at Halifax, Nova Scotia. Patents were subsequently obtained for the manufacture of kerosene oil by his process.

In England, in 1847, Mr. Charles B. Mansfield obtained patents for 'an improvement in the manufacture and purification of spiritous substances and oils applicable to the purposes of artificial light.' Mr. Mansfield's experiments seem to have been the base of the attempts to introduce the atmospheric light obtained from the use of the lighter oils or spirits, as benzole, in conjunction with atmospheric air.

In October, 1850, Mr. James Young, of Manchester, obtained a patent for the 'obtaining of paraffin oil, or oil containing paraffin, from bituminous coals,' and in March, 1852, Mr. Young obtained a similar patent for the United States. In defending these patents against manufacturers who were, by the processes they employed, supposed to be infringing the patents, much litigation was caused. Happily the result was that a great and rising industry was left unfettered, and no exclusive right was established in results which seem to have been gradually evolved by much thought and expense, and by many experiments, during a whole century. The distillation of oils from cannels and bituminous shales grew in the Flintshire coal-fields and in the coal-fields of America, as well as in most of the great cities of the Atlantic coast.

In 1853-4 kerosene oil of great lighting power but of unpleasant odour was introduced to the public in America by Messrs. J. H. and G. W. Austen, the agents of the North American Kerosene Gas-light Co. of New York. Safety and comfort in the use of this oil were further increased by the introduction into America by Mr. J. H. Austen of the adaptation of a burner which he had seen in use in Vienna.

In the course of the six following years the distillation of oil from natural petroleum almost entirely superseded the distillation from coal and shale. Petroleum had been collected from remote times in the State of New York by the Seneca Indians, and after them it had received the name of Seneca oil.

In the year 1854 petroleum was obtained from an old salt well at Tarentium, on the Alleghany, near Pittsburg, where its presence in quantity impeded the manufacture of salt. This was finally introduced successfully in New York by Mr. A. C. Ferris, in the year 1857. In 1858 the first petroleum well was bored by Mr. E. L. Drake, at Titusville, on Oil Creek, in Pennsylvania, with a successful result. The growth of the industry was after this most marked and rapid.

In the year 1861, three years after the boring of the wells at Titusville, the production of petroleum oils in the United States amounted to 24,000 gallons per day.

In 1862	40,000	per	day.
,, 1863	70,000	,,	,,
,, 1864	87,000	,,	,,
,, 1865	91,000	,,	,,

In the year 1882 the daily production reached the great total of 2,145,000 gallons per day.

These amounts are exclusive of the oil produced in Canada.

We will now notice the manner in which these bituminou

substances occur in nature, and the localities from which they are obtained, beginning, as usual, with the British Islands.

Bituminous matter forms part of the composition of the greater part of our coal-seams, the exception in this country being found in the western end of the South Wales coal-field, and in portions of the coal-fields of Ireland, where it would seem as if the bituminous matter had been driven off subsequent to the deposition of the coal-seams, possibly by the heat evolved during the great disturbances which have so crumpled and broken the strata.

Bitumen is especially present in the variety of coal known as cannel. This form of coal is most prevalent in the lower yard coal of Denbighshire and Flintshire, and in the equivalent seam in the Lancashire coal-field, as well as in what may be the same seam in the Newcastle coal-field.

In the Flintshire coal-field a great deal of cannel coal and its associated shale was used for the production of paraffin oils and products before the influx of petroleum from America at

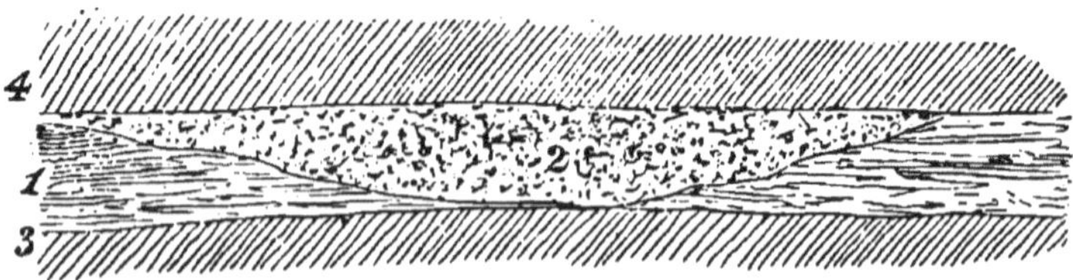

FIG. 51.—SECTION OF YARD COAL SEAM, FLINTSHIRE, SHOWING DEPOSIT OF CANNEL.

1, Coal Seam. 2, Deposit of Cannel in Coal Seam. 3, Bituminous Shale Floor. 4, Bituminous Shale Roof.

so cheap a rate made the manufacture unprofitable. The lower yard coal of the two counties of North Wales referred to is about a yard thick. The cannel form occupies but portions of the seam, filling up depressions in the ordinary coal, as shown in fig. 51.

In the cannel all vegetable fibre and organic structure, such as are seen in the other parts of the seam, are completely destroyed, and the coal appears as a hard pitchy mass. At such points also the under clay gives place to a bituminous shale, and in this and in the overlying shale of a similar character there are numerous fish remains, only indistinctly preserved. Where these coals and shales are in their perfect form, as about Leeswood, near Mold, the following is the order downwards :—

1. A rich oleaginous shale 4 to 10 inches thick.
2. Smooth cannel 2 feet 3 inches.
3. Curly cannel 1 foot 6 inches.
4. Floor of highly bituminous shale.

The smooth cannel (2) has a flat conchoidal fracture, and it passes downwards into the curly cannel. The curly cannel (No. 3) has a lustrous appearance. It is compact, and has a slightly conchoidal fracture. It abounds with flat circular disc-like appearances. It has some sulphide of iron, and shiny specks of arsenical pyrites. In the floor of bituminous shale, as well as in the roof shale, fish remains are abundant. Some years ago Dr. Andrew Fife, of Aberdeen, gave the following as the illuminating power of the various minerals named :—

	Cubic feet of Gas per ton.
Wigan cannel	12·01
Lesmahago coal	10·176
Torbane Hill mineral	15·482
Leeswood smooth cannel	9·972
Leeswood curly cannel	14·28

In places where the conditions have been favourable, the bituminous coals of Flintshire have passed again into a liquid form. Thus recently, in a colliery on Buckley Mountain, in Flintshire, a flow of oil was tapped in the workings.

In the neighbouring county of Shropshire, a bituminous spring has long been known in the parish of Wombridge, near Broseley. In the early part of the eighteenth century this spring is said to have yielded near a thousand gallons a week. In the year 1799 it only yielded about thirty gallons a week. At present it is not used commercially. The spring flowed from a fissured sandstone in the Coal-measures, having coals at a little distance above and below, and irregular patches of coal in it. An adit was driven in the sandstone, which facilitated the flow and collection of the oil. It is interesting to note that the water exuding from and near to the Flint coal—one of the lower coals of the Shropshire coal-field—is salt, and salt springs abound in the Coal-measures of the neighbourhood, some of which were formerly found of sufficient strength to work.

Perhaps the most interesting deposit of bituminous mineral in Great Britain, from its great commercial value, the part it has played in the development of the paraffin industry, and the litigation of which it has been the subject, is that of Torbane Hill, near the Firth of Forth, in Scotland. This deposit occurs in beds that lie below the ordinary coal-seams of the Lanark-

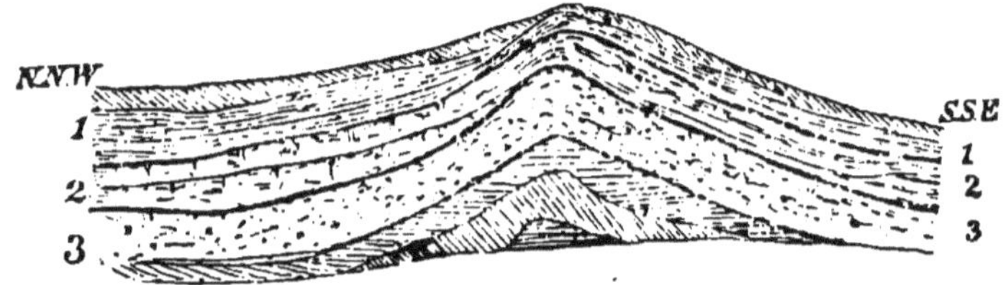

FIG. 52.—SECTION SHOWING THE POSITION OF THE TORBANE HILL MINERAL.
1, Lanarkshire Coal-measures. 2, Beds of Fresh-water Limestone on the N.N.W., with Shales and Coals on the S.S.E. 3, Shales, Sandstone, and Tufaceous Beds.

shire coal-field. Fig. 52 is a very general section of the strata of this field.

The Torbane Hill bed lies in group 2, on the north-north-west side of the section. It is associated with several coal-seams, and with them it occupies a small mineral basin about three square miles in area. These coal-seams and the particular bed itself are interstratified with two or three beds of fresh-water limestone. As will be seen, it underlies the ordinary Coal-measures. Bitumen occurs in a solid form in the underlying sandstone beds, and in round nodules in the limestones, as well as in contiguous trap' rocks, from which it also oozes out in a liquid form.

The mineral contains	volatile matter .	. . .	70·10
,, ,,	carbon or coke .	. . .	10·30
,, ,,	ash .	. . . . .	19·60

It yields 120 gallons of crude oil per ton. This, with the

exception of the bitumen of Ritchie County, Virginia, which gives 170 gallons, and the Breckenridge coal of America, which yields 130 gallons, is the largest quantity of oil yielded by bituminous substances. This mineral has been very largely used for the production of paraffin and its products. It was used under a patent by Mr. Young in the year 1850, and Young's paraffin oil, which gave the start to the use of mineral oil lamps in this country, was made from it. It was also extensively exported to the United States of America, and was used there in the manufacture of kerosene, in the production of which 200,000 tons were used in the year 1859 by the North American Kerosene Gas-Light Company, at their works, Newtown Creek, Long Island. The manufacture of oil from this mineral led to a vast amount of litigation; part of this lay between Mr. Young and the distillers of oil from cannel in England and the producers of petroleum in America, which happily ended in unrestricted manufacture, and part with the owners of the ground. The great question was whether the mineral was coal or not coal. The bulk of the evidence went to prove that it was not coal, but a bituminous shale or clay. Compromises resulted, and the use of the name coal was retained, although, as it has been suggested, the name might as well be given to the bitumen of the Great Pitch Lake of Trinidad.

A deposit of bituminous coal occurs, and has at various times been worked, near Bovey Tracey, in Devonshire. The deposit occurs in clays and sands that overlie the Greensand, and is of Tertiary age. The beds are from 4 to 16 feet thick, and are interstratified with the clays. It contains 42 per cent. of volatile matter, 58 per cent. of coke, and it yields 50 gallons of crude oil per ton. It abounds in the remains of fish, crustacea, and other marine organisms, and hence, like the crude oil of Canada derived from strata containing similar remains, it has a very strong smell. The oil contains a greater number of the equivalents of carbon than do the oils derived from coals or ordinary bitumen, and hence it is found to smoke in ordinary lamps.

It seems to have been first raised for use about the year 1700, and was subsequently being worked by means of pits about 80 feet deep. It is recorded that the offensive smell emitted by the mineral prevented its use for domestic purposes except by poor cottagers of the neighbourhood. It ceased to be worked about the beginning of the present century; but when the demand for oil shales sprang up, consequent upon the introduction of paraffin, and preceding the general use of American petroleum, a brief period of mining activity ensued; but the unpleasant smell and the density of the oil rendered permanent success impossible. It has, however, been used at various times in potteries that have been established for working the adjacent clays.

In Ireland, bituminous matter is occasionally seen oozing out of Silurian strata. More important was the attempt to manufacture paraffin from peat, which, during the demand for oil-producing substances, was made on an extensive scale by the Irish Peat Company in Kildare. As the use of peat may again be resorted to when the bulk of the petroleum wells have failed, it is well to place on record the results obtained by this company from one ton of peat.

Liquids not oily	65 galls.
Tar	6 „

From which were obtained—

Lamp oil	2 „
Lubricating oil	1 gall.
Paraffin	3 lbs.
Ammonia	3 „
Acetic acid	5½ „
Naphtha	8 „

with 25 per cent. of charcoal.

SWEDEN.—Bituminous matter, of more interest geologically than commercially, occurs at Millaberg, Wermland, Sweden. As shown in fig. 53, it occurs in gneissic rocks, and derives its interest from the fact that it shows the presence of vegetable matter in some force in strata of Laurentian age.

FRANCE.—At Feymoreau, near Fontenay le Comte, in the

Bourbon Vendée, is a small coal-field that stretches between Nantes and Rochelle, from some of the strata of which, as I have already stated, paraffin oils and products have been obtained for many years. The strata from which these products were, and it may be are still, obtained underlie a coal of inferior quality. They do not contain any vegetable impressions, which, we have seen, is usually the case where coal passes into a highly bituminous state, as in cannel. The shales are of a

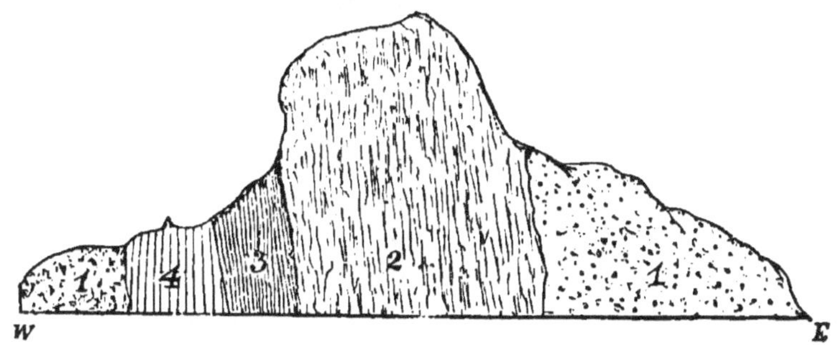

FIG. 53.—SECTION AT MILLABERG, WERMLAND, SWEDEN.

1, Reddish Granitoid Gneiss. 2, Gneiss. 3, Fine-grained Gneiss and Mica Schist. 4, Thin Strata of Bituminous Schist.

deep black colour, and when first mined are tough and hard, but they soon fall to pieces on exposure to the atmosphere. They burn freely, with much smoke and a long flame. They are not, however, uniformly bituminous. In places these shales reach a thickness of from 30 to 40 feet. Ordinarily they yield about 15 per cent. of light oil, on slow distillation, with 60 per cent. of ash and some water. They belong to the ordinary Coal-measures.

On the other side of France, about Autun, is another small coal-field of ordinary Coal-measure age. Here bituminous schists, like those of Feymoreau, occur much higher up on the series, and several hundred yards above the highest coal-seam. At Cordesse these schists are about 10 feet thick. They are of a reddish brown colour, but become black on exposure to the atmosphere. Some of the beds are pyritous; the quantity of oils of all kinds yielded by them varies from 6 to 50 per cent. They are worked partly by open quarrying and partly by drifts or adit levels, as is common in the coal mines of South Wales. The manufacture of paraffin, paraffin oils, and other products, has been carried on here since the year 1835 with varying success, the industry here, as well as in England, having been greatly affected of late years by the exportation of good oils at very low prices by America.

Of very great importance commercially in another direction—that of the construction of roadways—are the bituminous deposits of the banks of the Rhone, near Bellegarde.[1] These, with some of the uses of the bitumen, seem to have been discovered in the year 1721 by Dr. Eyrinis, who thought that in the bitumen he had discovered a panacea for all the diseases to which human nature is liable. It is on record, however, that from time immemorial the inhabitants have noticed bitumen; hence the name of the principal hill—Pyrmont. In the fifth year of the Republic a gentleman obtained a concession from the Directorate, his idea being to work the mineral for home consumption.

The concession extended along the two banks of the Rhone from Bellegarde to Seyssel—4 kilometres. It is from the last-named place that the chief mine of the district takes its name.

When the application of bitumen mixed with chalk began to extend, Secretan, the lessee, opened a mine and established a manufactory for the production of mastic. At this juncture some of the inhabitants disputed his right, on the ground that the concession had been obtained for bitumen only, contending that the use of bituminous chalk is different from that of free bitumen, and that the working of a quarry was not under the same law as was the working of mines.

After much deliberation, the Council of Mines decided that the works at Seyssel constituted a mine and not a quarry, and that the concession included the right to extract chalk as well as bitumen.

In the year 1843 these rights were finally confirmed, and in the meantime the mine continued to be the chief source of

[1] L. Malo, *Fabrication de l'Asphalte et des Bitumes*.

mastic. The output was, however, restricted until about the year 1838, when the use of asphalte in the construction of footpaths gave increased importance to and extended the operations of the mine.

In the year 1855 the production had risen to 1,500 tons yearly. In 1863 it had increased to 10,000 tons, and the production continued large subsequently.

The deposit at Seyssel consisted of a hill 400 metres long by 100 metres wide. It consists of chalk surmounted by beds of Greensand.

The chalk is made up of three layers of bituminous chalk from 3 to 4 metres thick, which are separated by three layers of white chalk not impregnated with bitumen, and which vary from 1 to 15 metres in thickness.

In the impregnated layers the chalk is in places highly crystalline, and in others it is made up of shells, among which fishes' teeth are found. The bitumen occurs in cracks, cavities, and layers. It forms from 8 to 10 per cent. of the whole mass.

The theory has been suggested that this bituminous matter has been derived from carbonaceous deposits far below, which have been slowly distilled by volcanic or chemically produced heat, the vapour arising and subsequently condensing, filling the pores and interstices of these Upper Jurassic strata. There are small proportions of arsenic and of manganese present, but the purer the containing chalk is, the better is the result in the manufactured asphalte.

It is possible, however, that it may have been derived from carbonaceous matter that formed part of the overlying strata which have been subsequently denuded, but of which traces here and there remain.

The mineral is got from the beds in blocks of different sizes, by means of quarrying in open excavations, the amount of top rock to be removed not being very great. In winter the rock is hard like ordinary chalk, and the drill works freely, but in summer the bitumen softens, and the rock is elastic. This destroys the ordinary action of the powder, and picks and wedges have to be used. In some other deposits which are worked underground as ordinary mines, this difficulty does not occur. There are several smaller deposits scattered about the neighbourhood of Seyssel which have been more or less worked, but the description just given will apply generally to them all.

Bituminous springs and deposits abound over the plain of La Magne, in Auvergne, some only of which have been explored. The mineral is of two kinds, calcareous and sandy. It is of a reddish colour and a pronounced arsenical odour. The principal springs are at La Fontaine de Paix, on the road between Pont du Chateau and Clermont. They give some hundreds of kilogrammes during the summer, but they are stopped during the winter.

Another important bituminous deposit, which is of the same geological age as that of Seyssel, is that of the Val de Travers, on the right bank of the Reuse. This deposit was also discovered by Dr. Eyrinis. It subsequently passed through several hands, and, with Seyssel, was made the base of extensive gambling transactions by speculators; but about the year 1860 it got into regular work, and is now an important source of the asphaltum of commerce.

Here the principal bituminous layer forms a lenticular mass of about 8 yards in thickness and 160 yards in breadth. It is covered immediately by an asphalt earth known as 'scrap,' and this is overlaid on the surface by a thin layer of vegetable matter. The deposit is richer in bitumen than those of Seyssel, and contains from 12 to 13 per cent. of bitumen. It is, or was until recently, worked as an open quarry.

GERMANY.—The slates below the brown coal of Tertiary age on the Westphalian side of the Rhine are very bituminous. They have been worked to a considerable extent, and the products have been manufactured at Beul, opposite Bonn. These shales are very thin, and they are known as 'blatte,' or paper-coal.

At Bamberg, in the north of Bavaria, and at Reutlingen, near Tübingen, in Wurtemburg, the Posidonia schists of the

Upper Lias have at times been worked for the manufacture of burning and lubricating oils and paraffin.

The same schists have also been worked at Orawicza, in Hungary.

By far the most important bituminous district in Germany is that of Hanover. This oil region extends from the city of Hanover, where oil is found in the suburb of Linmer, to the villages of Oilper and Klein Scheppenstett on the east and the Hildesheim Hills on the south; and it has been thought probable that the same condition of the strata may be found westward towards Bremen.

The strata are apparently of Permian age, and it is possible that, in parts of the area, they may have connection with and be influenced by the overlying Eisleben shales in which the copper slate is found, and in which there is an abundance of fish remains and those of other sea organisms.

The wells vary in thickness from 50 to 280 feet, according to the thickness of the strata overlying the oil-bearing beds. The general order of the strata is as follows, in descending order:—

	Feet.
1. Fine sand with boulders of granite and flint .	32
2. Bluish grey clay	23
3. Blue clay with layers of limestone . . .	10
4. Marl	65 to 80
5. Rock with veins of quartz	115 to 130
6. Sandstone with pyrites, showing the first traces of oil	Thickness not given.
7. Sandstone with black and red sand . .	
8. Layer of pebbles, in which the oil is found most abundantly	

In the neighbourhood of Bentheim, to the west of Hanover, in sinking to the oil-bearing bed a deposit of bitumen has been found. This mineral is found solid in the fissures of the sandstone. About 400 tons of this bitumen have been raised here from a mine belonging to Mr. Sargent. This mineral, distilled at the mine, yielded 110 gallons of petroleum to the ton.

This partially solid bitumen is frequently met with elsewhere in the same position in sinking shafts. It varies in thickness from a few inches to 3 feet. Over the whole area bitumen is seen in places oozing out of the strata.

As pumped out of the wells the liquid is composed of one-third water containing 2 per cent. of saline matter, and two-thirds oil. This crude oil, when refined, gives 45 per cent. of best petroleum.

This oil region has been worked more or less for the last two hundred years. In the year 1872 Professor Harper, of Pennsylvania, made a survey of the district, and during the last twenty years many trials have been made by boring. At the close of the year 1881 there were thirty companies formed, one hundred derricks already erected, and about sixty bore-holes in the course of sinking. At that time the average yield per well was from 10 to 15 barrels per day. An area of 40 acres near Olheim, which, with Oilper and Oilberg, forms a great centre of the industry, yielded for a time from 400 to 500 barrels a day.

A great *furore* prevailed about this oil region, chiefly, it is to be feared, for speculative purposes, in the year 1882, and the proprietors of the land acquired an exaggerated notion of the value of their properties. They demanded from 60*l.* to 150*l.* per acre for the right of boring alone, with the necessity of the lessee purchasing one-half of the property as spoiled land. The agreements generally included the right to purchase the whole property within a given time.

Some of these concessions were secured at a price of 13*l.* per acre, and in parts more distant from proved oil wells the right to bore was secured at prices ranging from 1*l.* to 8*l.* per acre.

The wells were sunk by the Pennsylvanian rope boring apparatus at a rate of from 30 to 40 feet per day.

CHAPTER XIII.

CARBON—continued.

Bituminous Deposits of Spain, Italy, Roumania—Bitumen and Petroleum of the Caucasus and Caspian Regions—Bituminous Springs of the Valley of the Euphrates, of British Burmah, of the Punjab—Pitch Lake of Trinidad—Bituminous Springs of Barbadoes, Cuba, Venezuela —Bituminous Coal Deposits of North America—Geological Age of the various Petroleum-yielding Strata of North America—Particulars of the Oil—Mode of Sinking Wells—Bituminous Springs and Schists of South America—General Conclusions.

BITUMINOUS SUBSTANCES—continued.

IN SPAIN, at Maister, about ten miles east of Vittoria, there is an extensive layer of bituminous chalk, very fine-grained, compact, and regularly impregnated with bitumen. Unfortunately, it is situated at the bottom of an almost inaccessible gorge, but this natural difficulty will, no doubt, be overcome when the mineral is wanted.

In ITALY, the oil of Agrigentum was famous in very ancient times, and in Piedmont crude oil has been pumped recently at Riva Nazzano, near Voghera, and also in the valley of Cocco. In Greece, a well in one of the Ionian Islands has yielded petroleum for upwards of two thousand years.

Extensive beds of lignite occur in ROUMANIA, and bitumen is found in different localities. The latter mineral has been worked near Prahova and Bouzes, where amber is also found. In north-eastern Roumania there are numerous petroleum wells, and the oil-bearing strata are estimated to cover an area of about 856 square miles. The present annual production is computed as equal to about 300,000 barrels.

The production of petroleum and other bituminous substances is also large in Southern Russia, the productive strata between the Caucasus, the Caspian and Black Seas, being of very great extent. Apart from the natural supply, utilised to some degree by the natives, latterly systematic attempts have been made to obtain petroleum by pumping. Two wells have been sunk in the valley of the river Kuban, which flows into the Black Sea, and many wells have been sunk in the district near Baku, on the Caspian Sea. These wells are generally sunk to a depth of 300 feet, and are said to produce 28,000 barrels of crude petroleum. A large quantity of sand comes up with the oil. The refined oil is said to be equal to that of America. During the year 1882 the petroleum trade of this region was very much developed, and 5,000 vessels are said to have entered and left the port of Baku, which were chiefly employed in the petroleum trade.

The production of refined petroleum for the year 1883 was 206,000 tons, and the firm of Nobel Brothers manufacture in addition yearly 450,000 tons of *mazoot*, or liquid fuel, besides other products of petroleum.

Reference has already been made to the abundance of bituminous matter on the banks and in the waters of the Euphrates, and we are familiar with the Bible reference to its use in the construction of Noah's ark and the building of the Tower of Babel. Evidence remains that it was used in the buildings of Nineveh, and to-day the springs of the region supply the inhabitants with oil.

In India, petroleum, now usually known as Rangoon or Burmese oil, has from time immemorial been obtained from wells on the banks of the river Irawadi, on the north-east border of Bengal, adjoining, and to some extent in, Burmah. These wells, of which there are usually from 500 to 600, are situated between Prome and Ava. The strata consist of marls of Greensand age, which are interstratified with beds of lignite and bituminous matter, and these are overlaid by sand and gravel. The wells are sunk through these strata to a depth of about 200 feet, when petroleum is obtained, and also naphtha.

The wells are mostly situated on the left or Burmah side of the river, which, near Wetmasut, forms a cliff several miles long and eighty feet high. Near Pugan, a little higher up the river than Wetmasut, there is interstratified a dark bituminous shaly limestone, which, from its fossils, is found to be identical in age with the London clay. Fossil wood abounds in the strata near Wetmasut. Along the whole length of the oil region between Prome and Ava there crops up at intervals an older limestone, probably of Silurian age. The recent production of these wells is stated at 400,000 hogsheads, about 25,000,000 gallons.

There are a number of springs in the Punjab, the yield of which, in 1880, was 2,850 gallons.

In the West Indies prominent notice must be given to the celebrated Pitch Lake of Trinidad. This forms the head of the harbour of La Brae. The bituminous matter, of the consistency of thick treacle, flows out of the hillside. It hardens on exposure to the atmosphere, but the newer streams flow over the older and more hardened layers. The surface of the hardened layers forms undulations in which are pools of water that contain fish.

Beds of lignite and bituminous matter are interstratified in the cliffs of the shores, and the pitchy fluid oozes out and flows over the water as I have described.

The hard bitumen is of a grey colour, rather brittle. It is varied somewhat in quality, as it is mixed with the sand and other matter over which it flows.

It was from the bitumen of Trinidad that Dr. Gessner first obtained kerosene, and the result of several trials he made as to its oil-producing quality is as follows :—

Specific gravity	0·882
Crude oil	70 galls. per ton.
Refined oil	42 ,,
Lubricating oil	11 ,,

The strata appear to belong to the Upper Tertiary.

Besides the Pitch Lake of Trinidad, the whole region contains springs of petroleum and bituminous matter. In St. Andrew's parish, Barbadoes, there is a petroleum spring, the product of which has been sold as 'Barbadoes tar,' or 'green tar.'

Several springs issue from a serpentine rock at Guanabacoa, near Havana, from which different varieties of bitumen are obtained. These were formerly used for pitching or 'careening' the ships visiting the place. Near Cape de la Brea streams of naphtha issue from the mica slate and cover the sea for some distance.

In the eastern part of Cuba, between Holguin and Mayan, there are also springs of petroleum.

The same geological conditions prevail in VENEZUELA as in Trinidad, lignite and bituminous beds being interstratified with other beds of Pliocene age. Near the Rio Tara, on the surface of a sandbank, there are a number of cylindrical holes of different diameters, through which streams of petroleum mixed with hot water occasionally issue out with great noise, and to the extent of four gallons a minute. There are also places where inflammable gas escapes from the soil. In the neighbouring republic of Columbia, between Escuque and Bellijoque, bitumen is collected by the labourers, and by a rough process of distillation oil is obtained from it.

NORTH AMERICA.[1]—In describing the important bituminous resources of North America, I will first notice some of the principal deposits of the more solid varieties of the mineral, and then describe the conditions under which the more liquid varieties of naphtha and petroleum are found.

Among the first-named, the Albert coal, of Hillsborough, Albert County, New Brunswick, occupies a chief place. It is described as a vein nearly vertical, from one to sixteen feet in thickness, differing in several respects from an ordinary coal-seam. It is associated with rocks strongly impregnated with bitumen, from which its contents have probably been derived. The mineral is brilliant in appearance, it breaks with a conchoidal fracture, and is highly elastic. It dissolves in naphtha and melts in the flame of a candle. Its composition is as follows:—

Carbon	85·400
Hydrogen	9·200
Nitrogen	3·060
Oxygen	2·220
Ash	·120
	100·000

[1] Hitchcock, Gessner.

The average yield of crude oil, as shown by several trials, is 110 gallons per ton, or—

Volatile matters	61·050
Coke	30·650
Hygroscopic moisture . . .	0·860
Coke	7·440
	100·000

This mineral has been the subject of much litigation, hinging upon the question whether it was asphaltum or coal. Coal had been reserved by the Government, but no mention was made of asphaltum in the original grants of the land. In the spring of 1852, a Halifax jury decided that it was asphaltum, basing their decision on the scientific evidence laid before them. In the summer of the same year another jury, after a trial of eleven days, gave a verdict in favour of the mineral being coal. Perhaps, after all, it is only a vertical cannel coal seam.

Breckenridge Coal.—Among the coal beds of the vast Alleghany or Apalachian coal-field of the United States, there are several beds of cannel, including that which is known by the above name. This bed, which is a rich cannel, is worked in the county of Breckenridge, Kentucky. It is a bed of about three feet in thickness. It yields at a red heat—

Volatile matter	61·300
Fixed carbon	30·000
Ash	8·055
Hygroscopic moisture . . .	·645
Sulphur	trace
	100·000

The results are, however, variable, as is the quality of the coal. Ordinarily it yields 130 gallons of crude oil to the ton, of which 58 to 60 may be made into lamp oil, with varying proportions of the heavier oils and of paraffin.

Petroleum-bearing strata range over an immense area in North America, occupying large districts comprising several hundred thousand square miles, from Carolina in the south to Canada in the north, and from the extreme east to the extreme west. These strata are of almost every geological age, as the following particulars will show.

Tertiary Strata.—To the Pliocene division of this group belong the oil-producing rocks of California. In the Cretaceous strata of Colorado, Nevada, and the plains of the Mississippi River both petroleum and lignite beds occur. In North Carolina and along the Connecticut River oil occurs in beds of Triassic age. In Western Virginia the bulk of the oil wells are in strata that lie near the summit of the Coal-measures. Lower down in the same series, near the Pittsburg coal, are the oil wells worked at Wheeling and Athens. Four hundred and seventy-five feet lower down still, near the Pomeroy coal-bed, is another oleiferous band; and again oil is derived from near the base of the Coal-measures in the top beds of the Millstone grit. The Archimedes limestone, which lies near the base of the Carboniferous limestone, has produced oil in Kentucky.

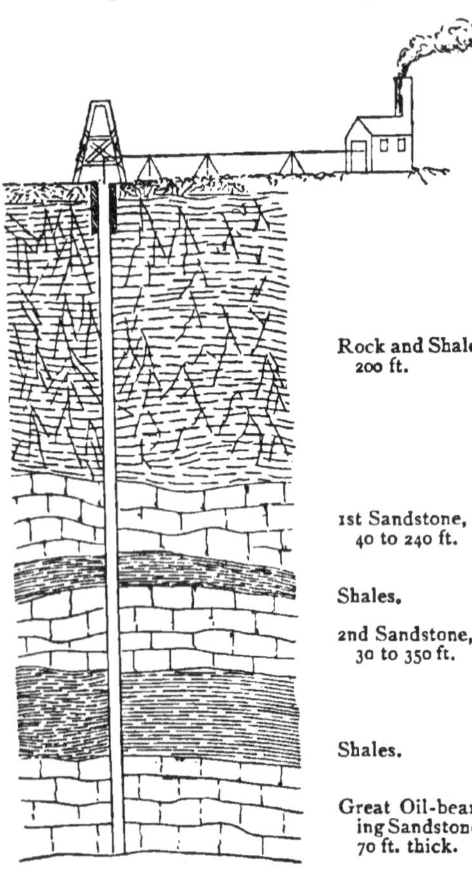

Fig. 54.—Section of Oil-bearing Strata in Pennsylvania.

In the Chemang and Portage group, which corresponds to our Middle Devonian strata, are the petroleum wells of the West Pennsylvanian oil region, as well as those in north-east Ohio. These strata have a combined thickness of from 2,000 to 2,500 feet. At the mouth of Oil Creek the lowest beds of the Coal-measures are seen capping the hills. The intervening rock between these and the uppermost strata of the valleys form the hillsides and the uppermost strata of the valleys.

In these valleys the strata intimately connected with the oil-bearing rocks are shown in fig. 54.

The average depth of the wells is about 650 feet, but wells have been sunk to a depth of 1,200 feet. The depth depends upon the amount of strata overlying the first sand rock at any particular point, and also, in a lesser degree, upon the variation in the thickness of the sandstones and shales down to the third sandstone. Oil is usually struck in the first sand rock, and it occurs more or less in the strata down to the third sandstone, but it is in this rock that is the great oil-producing bed of the region. The separate beds of sandstone are readily distinguished by their enclosed fossils.

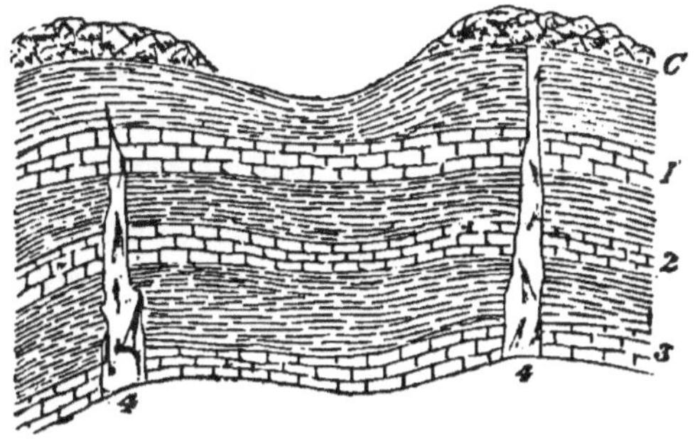

These strata are throughout the region thrown up in anticlinal ridges and synclinal troughs, as shown in fig. 55.

FIG. 55.—SHOWING UNDULATIONS IN THE OIL STRATA OF PENNSYLVANIA.

C, Base of Coal-measures. 1, 2 and 3, Oil-bearing Sandstone Rock. 4 4, Cracks occurring in the Anticlinal Ridges.

The bulk of the productive oil wells have been sunk in the valleys, where the oil has accumulated in the troughs of the strata. But productive wells have also been worked in parts of the ridges where, in the process of upheaval, the strata have been broken, forming cracks of various magnitude.

Authorities are divided in opinion as to the age of the Canadian oil strata, some making them the equivalent of our

Wenlock limestone and shale, while others place them near the base of the Cambro or Lower Silurian. Possibly in different localities both of these groups of rocks contain pretroleum. One characteristic of the Canadian oil strata is the profusion of molluscan and crustacean remains they contain, and it is from these, as well as from the remains of other marine vegetation associated with them, that the oil is supposed to have been derived.

Some of the natural oil springs about Dereham and Enniskillen occur along the line of a long, low anticlinal ridge which runs east and west, and is made up of a limestone with overlying shales. Springs also abound along the Thames for sixty miles west of Dereham. The oil seems to have been accumulated in the cavities of the limestone, and it is probably derived from the overlying shales with their vegetable and animal remains. In places the shales and the limestone are overlaid by drift ranging from forty to fifty feet in thickness. Productive oil wells have been sunk in similar strata in Kentucky and Ohio. Canadian petroleum was known to the early settlers, and on Black Creek pretroleum springs had covered about two acres of land. These have left solid bitumen on the surface, the lighter portions having evaporated.

The quality of the oil varies according to the nature of the strata from which it is obtained, and, as in the great Pennsylvanian oil region, according to the depth from which it is worked. Heavy oils rich in the various ordinary products come from the shallow wells, the lightest oils coming from the greatest depths.

The oil from the Pennsylvanian wells is of a greenish colour and in a crude state of an unpleasant smell. These oils yield when distilled 70 to 85 per cent. of a very fine burning oil, which is usually sold subject to the test that it will not inflame or pass into vapour under a temperature of 108 to 116.

Some of the wells in Pennsylvania and some in Ohio give heavy oils of a dark amber colour which yields about 90 per cent. of petroleum and 5 per cent. of naphtha, and which is used in woollen manufacture instead of lard oil. Some of the

Ohio oils make excellent lubricators, and about Duck Creek fine burning oil of a light colour is obtained.

Two tests of petroleum from California gave the following results :—

1.		2.	
Burning oil . . .	38 per cent.	Burning oil . . .	50 per cent.
Lubricating oil . .	48 ,,	Light lubricating oil	20 ,,
Pitch	10 ,,	Naphtha	5 ,,
Water	4 ,,	Heavy oil and paraffin	25 ,,

Crude oil from the Buonaventure district in California give 50 per cent. of burning oil and 28 per cent. of lubricating oil.

The Canadian crude oil is of a dark colour and offensive smell, which was at first very much against its sale. With the improved methods of distillation now in use this smell may now be effectually removed.

In the principal oil districts of the United States, on lands containing or supposed to contain oil, the right to raise the oil is usually leased by the owner for a period of thirty years. He reserves to himself a royalty ranging from one-tenth to one-half the oil, and he frequently receives at the outset a bonus of from 1,000*l.* to 2,000*l.* The lessee is bound to commence work within two months, failing which he forfeits the lease, as he also does if he continues work beyond a specified time. Occasionally a lump sum is paid down for oil rights at the beginning. In Western Virginia this ranges from 40*l.* to 200*l.* per acre. In the heart of the Pennsylvanian oil region the price is 200*l.* per acre. Further from the centres of the oil industry the price ranges as low as from 20*l.* to 50*l.* per acre.

The wells are generally sunk with ordinary boring tackle worked under a square-framed derrick forty feet high. A pipe, usually 4 inches diameter, is first driven through the soil and jointed as it goes down, to a depth frequently of thirty feet. It is cleared of its contents of earth, clay, and sand by a suitable tool, and then the boring rods are put to work their way through the rock like the old-fashioned boring for coals in this country. Before the hole is deep enough the whole of the boring tools usually weigh half a ton. When the hole has been taken down to the required depth, it is lined by a two-inch wrought-iron tube. In the first part of this tube that is let down the valves are carefully placed, and the first length of the pumping rods is attached. The remainder of the pipes are then screwed on in 12-feet lengths, and the pump rods attached as the tube is let down until the whole length reaches the bottom of the hole.

The wells were sunk by this, the old arrangement, at the rate of six to eight feet a day, and the total cost of a well about sixty feet deep would be about 2*l.* 5*s.* a foot. Drilling machines are now used with advantage as to the saving of time, but probably at not much less cost than formerly.

Care is taken to prevent water flowing down to the pump by placing a bag containing linseed, so made as to encircle the tube at a given depth. The seed on becoming moist swells and fills the bore-hole, and effectually prevents the downflow of superficial water to the oil.

SOUTH AMERICA.—Bituminous schists occur near Mendoza, in the Argentine Republic, and liquid petroleum springs exist south of the city. An important deposit of petroleum also occurs about 200 miles from Mendoza on the road leading to the 'Planchon' Pass for Chili. The crude petroleum here yields 40 per cent. of pure kerosene oil.

The liquid is discharged from fissures and other apertures in the rock. In the summer it flows more quickly and to a greater distance than in the winter. It gradually cools as it flows to a distance, and becomes a hard, compact mass of bitumen.

From the foregoing description of the geological conditions under which bituminous substances occur in nature it will be seen—

1. That they are more or less connected with deposits of vegetable matter, either forming part of such deposits, as in the case of cannel coal, the Boghead, Autun, Breckenridge, and other similar deposits, or derived from them, as in the case of the French and Swiss bituminous calcareous deposit, the Indian and American petroleum wells, and the Pitch Lake of

Trinidad, and the bituminous matter of the Caspian Sea and the regions of the Euphrates.

2. Even where the original deposits of vegetable remains are not now seen overlying the bituminous matter, we may now assume, from slight indications that remain, as well as from the analogy of similar deposits elsewhere, that they once did and that they have probably been subsequently removed by denudation.

3. We see that where bituminous matter abounds in a bed of vegetable remains, as in the cannels of Flintshire, the conditions under which such bed was originally formed were different from those under which ordinary coal beds were deposited. From the entire destruction of vegetable structure, and the presence above and below of fish remains, we infer that the vegetable matter of which bituminous coals were formed was deposited in water, and—from the frequent issue of salt water from such deposits—in salt water, probably in shallow lakes bordering upon the sea, and that they were composed of plants of different sizes and structure to those forming the less bituminous beds.

4. We infer that so deposited in lagoons and marshy places, the juices of the plants themselves would escape evaporation, and also that in the process of decomposition the carbon of the vegetable matter would absorb and become chemically mixed with a portion of the hydrogen of the water by a similar process to that we see occurring in hay, straw, and other substances put together in a wet state, and partially excluded from the atmosphere, and in the course of which much heat is evolved.

5. As these deposits became covered up and compressed into a completely solid state, and to the entire exclusion of air and further moisture, their chemical condition became fixed, as we see in the case of coal-seams when first struck underground.

6. When, consequent upon the breaking up of the strata by processes of upheaval and depression, air and moisture have been again admitted by cracks and by exposure in hillsides

and cliffs, chemical changes gradually follow, which of themselves probably give off heat enough to cause a slow distillation to take place. This result is illustrated in the issue of petroleum from the cannel coal-seams in the Buckley Mountain colliery, in the petroleum spring from the Shropshire coal seams at Madely, and in the flow of liquid bitumen at Trinidad.

7. We are thus led to the conclusion that all the forms of liquid bitumen have been derived by a slow natural process of distillation at a low heat produced in the mass by chemical action; that, favoured by natural fissures, the liquid has flown into cavities and natural receptacles in strata that are of older date than the source whence the liquid was derived.

8. The chemical changes referred to have been accompanied by a large production of inflammable gas, which in many of the American wells has been the chief agent in forcing the liquid volume to the surface. The water by which the oil is accompanied is that which has found its way from the surface, and derives its frequent saltness from the strata through which it has flowed.

AMBER.

Amber was called *electron* by the Greeks, because it became electric very readily when rubbed. It was also called *succinum*, or juice, from its supposed vegetable origin.

Its chemical composition is: carbon 72·0, hydrogen 10·5, oxygen 10·5. It is yellowish in colour, ranging to white and brown, and is transparent to translucent. It occurs in clays chiefly of Tertiary age in France; in the London clay of the Paris basin; in Austria on the coast of the Adriatic, and in Poland. It is most abundant on the German coast of the Baltic, in the peninsula of Samland, between the bays Tusche Haf and Kurische Haf, and especially near the town of Könisberg. It is imbedded in clay of Tertiary age, out of which it is washed by the sea where the clay is within the action of its waves, and it is collected on the sea-shore. It is also obtained direct from the clay by digging and subsequent washing and sifting. The industry gives employment to about 1,400 persons, and about 130 tons a year are collected. The right of gathering the deposits in the lagoons of Tusche Haf and Kurische Haf has just been let by the Prussian Government for a further term of twelve years to the firm which has enjoyed the monopoly for the last twenty-four years, for the sum of 120,000 marks. It is estimated that each cubic foot of earth contains from half a pound to a pound of amber. Amber frequently contains the remains of insects which got entangled in it when it was a viscous fluid, and many signs of their struggles to get free are apparent. The earth in which the amber is found seems to have been derived from the waste of Greensand beds, where these rested on an old Silurian rock, and it is supposed to have come into its present position by the wasting of such rocks in Scandinavia and Finland.

The trees are such as grow in temperate zones—camphor-trees, willows, beeches, birches, and oaks, with pines and firs, including the amber-pines. To have furnished the quantity of amber found it is supposed that many thousands of amber-pines must have perished and the amber gum have accumulated in the soil before the submergence of the land took place.

CHAPTER XIV.

SULPHUR.

Abundance of Sulphur in Nature—Varieties—Sulphur Mines of Sicily—Situation—Geological Position—Details of Strata—Thickness of Sulphur Beds—Percentage of Sulphur contained—Associated Minerals—Methods of Working, Costs—Sulphur Mines of the Mainland of Italy—Cessena—Geological Position—Thickness of Bed—Modes of Working and of the Treatment of the Mineral—Sulphur Deposits of Greece, of Russia, of Iceland—PYRITES, Production of in the British Isles—Growth—Of the Treatment of on the Tyne—Pyrites of Norway, of Germany, of Spain and Portugal—Description of the Rio Tinto Mines of Spain.

SULPHUR is an abundant mineral in nature. In a finely disseminated form it is present in most rocks. It enters largely into the contents of mineral lodes. It is present in force in particular beds in different geological formations, and extensively associated with the ores of most metalliferous minerals—silver, copper, tin, lead, zinc, iron, and many others. With iron and copper it is present in such abundance that the sulphuret of these ores forms one great source whence the sulphur of commerce is derived. It also occurs native in the manner and under the conditions hereafter described.

Native sulphur is in colour and streak yellow, sometimes of an orange yellow, with a resinous lustre, and transparent to translucent. It is soft, and may be scratched with the nail. The specific gravity is 2·07. It occurs perfectly pure and also mixed with clay, bituminous and other substances. When it contains selenium, it is of an orange yellow colour. It burns with a blue flame and a sulphur odour. It is a very useful mineral, being largely used in the manufacture of gunpowder, sulphuric acid, and for bleaching and medicinal purposes.

The chief deposits of native sulphur are found in Italy. Indeed, up to the present time this country has possessed the almost exclusive monopoly of the production of sulphur, although there are lesser deposits of the mineral in France, Spain, Greece, and on the borders of the Red Sea.

The chain of the Apennine Mountains, which runs down Italy and is continued into Sicily, is made up of Tertiary strata, and at various points along its course these strata furnish indications of the mineral, which is to some extent mined. But the bulk of the sulphur produced by Italy is obtained from the deposits in Sicily,[1] the annual production being estimated at 200,000 tons. The Government mining returns for 1881 estimates the value of the whole of the sulphur mined in the kingdom at 1,000,000*l.* In Sicily the central chain of mountains stretches from Messina on the east-north-east to Marsala on the west-south-west. Another chain crosses the island from Nito, on the south-east, and from the main chain about midway in its course opposite Cefalu. In the triangular space formed to the east of these two ranges rises the isolated volcanic mass of Etna. A little basin of strata containing sulphur is found to the north of the main chain in the province of Palermo. With this exception all the sulphur deposits of the island are found south of the main chain, and on both sides of the south-east and north-west chain, but chiefly on its west side. There are about twenty points at which the deposits have been proved and worked, the principal mines being situated near Cattolica, Girgenti, Licata, Caltanisetta, Centorbi, and Sommatino.

Stratigraphically the sulphur deposits seem to belong to the middle of the Tertiary division of strata. The following is the general succession of these strata in the district, in descending order :—

[1] 'Les Mines de Soufre de Sicile,' par M. Ch. Lectorix, *Annales des Mines*, série 7, tome 7.

Pliocene.	1. Sandstone and calcareous cement, conglomerated sands with intercalated layers of fossiliferous marls and gypsum beds, thickness 100 to 150 metres.
	2. Calcareous tufa, chiefly made up of fossils in little accumulations, 100 metres.
	3. Bluish grey marls, 38 metres.
Upper Miocene.	*4. White calcareous marls with foraminifera, the colour sometimes being grey, 50 to 120 metres.
	*5. Saccharoid gypsum, crystalline gypsum, and foliated gypsum, with impressions of fish remains with sulphur beds, 20 metres.
	*6. Calcareous sulphurous marls, tufas, and gypsums, 35 metres.
	7. Compact calcareous marls, sometimes silicious, corresponding to the bed called *cagnino* in the sulphur mines of the Romagna, 1 to 30 metres.
	*8. Tripoli, divided sometimes by a deposit of calcareous and magnesian marl, thickness variable.
Middle Miocene.	9. Quartzose and micaceous sandstones with slightly saline marls and intercalated conglomerates, formed partly of fragments of crystalline rock and partly of ferruginous sandstones, 10 to 40 metres.
Lower Miocene.	10. Deposits of salt crystals, generally missing where No. 9 is present in force.
	11. Saliferous and gypsiferous marls of a bluish colour, containing petroleum and bituminous substances, 600 to 1,000 metres.
	*12. Calcareous concretions with silica, 15 metres.
	13. Ferruginous and gypsiferous clay with bituminous schists containing arragonite, 1,500 to 2,000 metres. This clay, over the northern part of the sulphurous zone, contains a limestone perforated by nummulites.
Eocene.	14. Nummulitic limestone containing fucoids and jasper.

The order of superposition given above is not very constant, some of the members of the series being frequently missing. Thus near Caltanisetta the silicious limestone and marls, No. 7 is missing, while at the other parts, notably at Grotacula, Ries, and Sommatino, it is strongly developed. At Casteltermini also the gypsums, No. 5, are found both above and below the sulphur deposit No. 6.

Of the foregoing strata, Nos. 4, 5, 6, 8 and 12 are those which contain the deposits of sulphur.

The sections, figs. 56 and 57, show the particular grouping of the sulphurous strata near Caltanisetta and Sommatino, the section of the beds at the latter place showing also the highly disturbed nature of the strata—a condition so prevalent that it is a local notion that highly inclined disturbed strata are the most productive of minerals.

The sulphur occurs in the. beds in masses, lying in basin-shaped depressions in the sulphurous beds. These are separate from each other. They average in size 20 kilometres in length by 3 kilometres in breadth. The mineral is disseminated in the mass of these basins in nests and in irregular

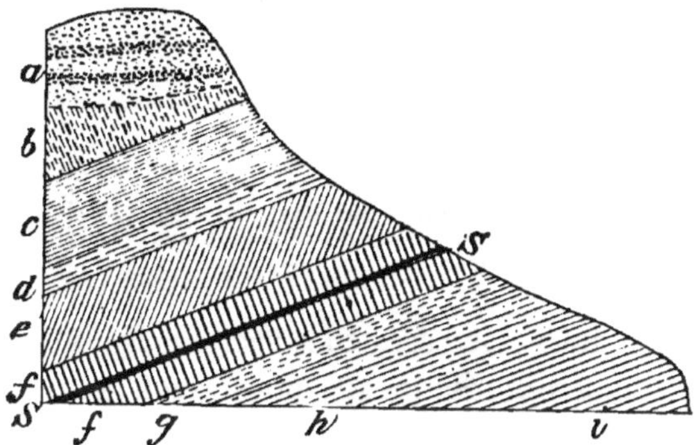

FIG. 56.—SECTION OF THE SULPHUROUS STRATA, NEAR CALTANISETTA, SICILY.

a, Grits and Yellow Sands. *b*, Calcareous Tufa. *c*, Veins of bluish Clay. *d*, White Marls, foraminiferous. *ef*, Clays with Gypsum. *g*, White Marls with Fish Impressions. *h*, Sulphurous Clay. *i*, Grey Clay. S S, Sulphur Bed.

branching veins, and in little beds lying parallel to the stratification; exceptionally, as at Racalmuto, the sulphur lies in a sand dyke, cemented by both calcareous and bituminous matter. The sulphur is ordinarily amorphous, but it is met with frequently in flat, round, and oval-shaped masses that contain beautiful eight-sided crystals. Its ordinary colour is yellowish brown, with a resinous aspect. Sometimes it is yellow, with a slight tinge of green. There is also a soapy variety that is opaque and of an ochreous nature. The associated minerals are sulphate of lime, carbonate of lime, and more rarely, sulphate of barytes. The deposits of sulphate of lime (gypsum), are in intimate relation to those of the sulphur, and are apparently due to the same causes. This mineral is met with as a roof to the sulphur beds, and it also occurs in separate masses in the midst of them. It is also largely present in the marl beds, where it takes the name of balatino. The thickness and number of the sulphur beds or deposits in any one of the foregoing divisions is very variable. At the

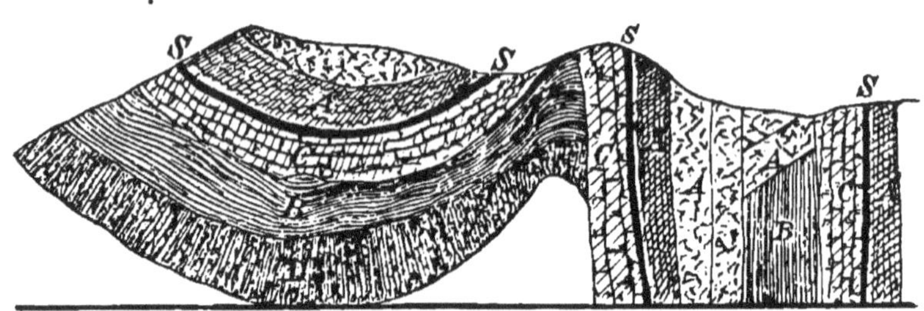

FIG. 57.—SECTION EAST AND WEST THROUGH THE HILL SCORSONE, NEAR SOMMATINO, SICILY.

A, Gypsum in large Crystals. A′, Gypseous Marl. B, Tripoli Marls with Fish Impressions. C, Calcareous Marls and Limestones with Gypsum. D, Saliferous Clays with Salt Crystals. S, Sulphur Beds.

mine Madore, near Lercara, the section is as follows, and the beds are called by the local names given :—

Bed.	Metres.		Metres.
Zagareddada, ribbon-like deposits of sulphur	8 0 0		
Blackish schisty clay	1 5 0	to	2 0 0
Percullatella, nests of sulphur in limestone	2 0 0	,,	2 5 0
Clay	2 0 0	,,	4 0 0
Percullatella grandi	4 0 0	,,	6 0 0
Clay	5 0 0	,,	6 0 0
Orlando	5 0 0	,,	6 0 0

or giving from 16 to 22 metres of sulphur.

At Caltanisetta the average thickness is 4 metres. At Grotacula 12 metres in three deposits, which are separated by clay of from 1 to 1½ metres thick. At the great mine of Sommatino the thickness reaches 30 to 33 metres, and the deposit is divided into six parts of from 2 to 8 metres, which are separated from each other by barren partings of about 1 metre thick. At Lercara the deposit reaches a thickness of 33 metres without any barren partings at all.

The minimum thickness of the beds is from 1½ to 2 metres. They are extremely disturbed, inclined, and distorted, and thus different portions of the same deposits are separated from each other.

The percentage of sulphur contained in the mass of the deposits is also very variable. At Madore it is 20 per cent.; at Grotacalda, 25 to 27; at Sommatino, from 18 to 26; at Cimico, near Racalmuto, 21. In some mines it is as low as 12 per cent. Sometimes near the surface hydrogen is intimately mixed with the sulphur, and this is considered a good indication. When this sulphur is exposed to the air it decomposes, and becomes similar to the soapy sulphur that is found elsewhere in the deposits.

The underlying clays or marls are nearly always bituminous, and occasionally, as at Racalmuto, the sulphur itself is impregnated with bitumen. Hydro-carbon emanations are still continued, and are noticeable at many points, more particularly near Caltanisetta and Girgenti, where there is volcanic mud, salt water, and hydro-carbon emissions.

It has been assumed that these sulphur deposits were formed in a series of lakes or lagoons, containing fresh or brackish water, into which sulphate of lime found its way, the sulphur being thrown down by volcanic emanations, and the organic matter permeating the mass of the underlying clays.

Associated with the sulphur deposits, or contiguous to them, are those of saltpetre, chloride of potassium, chloride of magnesia, and sulphate of soda. The range of the salt deposits commences to the south of Nicosia, and extends to near Cattolica. The greatest breadth of this saline band is about 20, and its length 120, kilometres. The salines do not constitute a continuous deposit over all this area, but they are concentrated in different groups, the most important of which are those of Leonforti, Priolo, Cranara, and Trabona, between Caltanisetta and Marianopolis. In these deposits there are also the chlorides of soda and magnesia.

The working of the Sicilian sulphur mines is for the most part in a very crude state. Until recently the miners were ignorant of the use of vertical shafts and of the use of timber for supporting roofs. They knew little or nothing of a plan, nor of any engines, machines, or mechanical appliances for the extraction of the minerals or the drainage of the mines. This state of things has resulted from the absence of mining legislation, from the profound ignorance of the people, the isolation in which they always live, the want of communication into the interior of the country, the want of fuel, vegetable or mineral, and the scarcity of wood for mining purposes.

Of late years strangers have embarked in mining operations on the island, and they have endeavoured to introduce improved methods of mining. But they have met with great opposition from the lack of resources in the country itself, and also, and most of all, from the opposition, latent and open, of the people of the country. This resistance, organised in secret, finds means of manifesting itself against modern innovations that arouse their antipathies. The consequence is that it is computed that at nearly all the three hundred and fifty mineral works not more than a quarter of the work is done that ought to be done for the money expended. In the mines worked by strangers a change for the better is apparent, and, with the extension of railways, the development of instruction, and the efforts the Italian Government is making to repress crimes of violence, and in the deepening of mines, which render the ancient methods of working useless, there are signs of improvement, but it is felt that the progress is slow, and the full improvement desired will not come yet.

When the bed of sulphur crops to the surface itself, and the inclination is not more than 45 degrees, a descent is made in the bed itself. If the sulphur itself does not crop up to day, but the *buscale* or containing bed is there, or the patches of sulphur in the wall of limestone (*calcare du mur*), or the soapy sulphur presents itself, a *tentative* or inclined shaft (*buci* or *scalier*), which is designed so as to reach the deposit in the shortest distance possible, is made. This is also done when the inclination is more than 45 degrees. These inclined galleries are provided with simple ladders when their slope is from 30 degrees to 35 degrees. When the slope is more than this, two sets of ladders are placed side by side, and they are so arranged that the landing-place of the one is half way up the other.

The mineral is carried up these ladders on the backs of children and young people of from eight to eighteen years of age, called *careizi*, the weights ranging from 20 lbs. to 80 lbs., according to the age and strength of the bearer. They carry the big lumps directly on their shoulders. The small is carried in sacks made of rushes. It is not without danger that these long files of *careizi* race up and down the *buci* and ladders. Sometimes one of them lets fall his burden, to the hurt and discomfiture of those below.

When a little water is met with, it is drawn with bottles of baked earth that hold from 16 to 20 litres, and these are passed from hand to hand upwards. When the inflow of water is considerable the work is abandoned. A new one is usually commenced by the side of the old one, which usually meets with the same difficulties and fate, the workers only winning the sulphur that lies above the water-line.

The *buci*, or inclined shafts, when made in the gypsum or limestone, usually stand without timber, but when they traverse the marls they are maintained by a casing of stones cemented in plaster, which plaster is baked at a low temperature obtained by burning straw. The plaster sets rapidly, but on exposure to the atmosphere it falls to pieces and needs constant repairs. In the mining excavations in humid marls this casing is quite insufficient, and the whole work collapses and is abandoned.

When the *buci* reaches the deposit a number of openings are made in the latter, usually without any method except that of following the rich mineral, and leaving pillars to support the roof. The roof being usually of clay, a thickness of about a yard of the sulphur is left to support, and when the chambers are very large the sides are strengthened by walling, as already described. The mineral is extracted by a pick with a blunt point; powder is seldom used, especially in the rich mineral, its use being attended with danger. The pick-men make six hours actual work in the mine per day, and each man is supposed to extract about one ton. The men receive from two to three lire a day. The price paid for extraction depends a good deal upon the depth of the mine, the thickness of the bed, and the hardness of the enclosing rock. At the Madore mine, which is of an average depth of 45 metres, the price is 13 lire per cubic metre. At Raculmuto, with a depth of 65 metres, it is about 15 lire.

At the Madore mine the cost of a ton of ore is estimated as follows:—

	Lire.
Breaking down	1·57
Subterranean and shaft carriage	2·04
Lights, tools, and repairs	·37
Putting on heap at surface	·25
Keeping up the mine and works	·42
Drainage	·42
Superintendence and general expenses	·90
	5·79

At the mines of Cornica the cost per ton is estimated at 5·12. The average cost per ton is given at 5·38. To this sum has to be added the cost of general administration, taxes, redemption of capital, &c.=1·14, making a total cost per ton of 6·52.

We will take as an example of the sulphur mines worked on the mainland of Italy those grouped near the ancient episcopal town of Cessena, and which are worked with English capital and skill. These mines comprise several properties—the Boradella, the Ca di Guido, the Borella Tarna, the Polenta, and the Monte Codruzzo. These properties are situated on the slopes of the sub-Apennine chain of mountains. As in Sicily, the enclosing strata belong to the Miocene division of the Tertiary strata. The sulphur deposits occur in marls that contain a good deal of sulphate and carbonate of lime. The chief bed worked is from nine to twelve feet thick, and with the enclosing strata forms a shallow basin-shaped deposit. The mineral contains 22 per cent. of pure sulphur, the ore occurring in ribbon-like strings and masses of varying sizes.

The mineral is won by shafts, of which there are a good many. These are sunk to the marl that underlies the sulphur, one of them attaining the depth of 1,100 feet. From the bottom of the shafts levels are driven and galleries opened out in the deposit, pillars being left to support the roof. Not without considerable opposition the principal shaft has been furnished with modern pit-gear, and with guides and cages, all of which created serious consternation when they were first introduced.

When the sulphur mineral is brought to the surface it is placed in large circular kilns, like lime-kilns. Of these there are fifty-five conveniently placed in relation to the shafts; each kiln will hold 350 tons of mineral at one charge. When full the mineral is ignited, the burning sulphur furnishing the greater part of the heat required for treating itself. The brimstone falls into the basin at the bottom of the kiln, from which, at stated periods, it is run into blocks of brick-coloured sulphur called black sulphur. In the process of calcination a loss, estimated at 35 per cent., is sustained, owing to the escape of the portion ignited as sulphurous acid. Experiments are being made, with only a limited success hitherto, to prevent this loss.

The total cost of mining the ore, calcining and producing the black sulphur, is given at 3*l.* per ton, inclusive of delivery. The selling price is 5*l.* 5*s.*, showing a profit of 2*l.* 5*s.* But it is probable that when all outside and exploratory charges are provided for the cost will be considerably increased.

About one-seventh of the produce of the mines is sold as black sulphur; the rest is converted into pure sulphur by sublimation and condensation in iron retorts. During this refining process a further loss of about 3 per cent. is sustained. The refined sulphur sells on the spot for about 6*l.* 2*s.* 6*d.* per ton, the profit being estimated at 1*l.* 15*s.* to 1*l.* 17*s.* 6*d.* per ton. The mines are worked on the pillar and stall system. The difficulties to be contended with are the swelling of the strata, as in a coal mine, falls of roof, barren ground, and fires. A very disastrous fire occurred in the old workings in 1872. A few years since the annual production of sulphur from these mines was about 12,000 tons.

GREECE.—The sulphur deposits of Greece occur about three miles east of the Isthmus of Corinth, and they lie in a series of cream-coloured and grey gypseous marls that underlie a white Miocene limestone. They therefore occupy the same stratigraphical position and occur in similar mineralogical associations to those of Sicily. The limestone ridges rise to a height of 2,000 feet on the northern shore of the Sea of Ægina, and show a disturbed condition of the strata. They are broken by rocky gorges and fissures. At a height of about 250 feet above the sea the marls are loaded with sulphur. On entering these gorges and the caverns and fissures by which they are traversed, the temperature is found to be about 100°; large bodies of stifling vapours are emitted, and great heat is discernible on the floor of sulphur, the whole seeming to indicate a present connection with volcanic heat. The caverns are completely lined with crystals of sulphur and other minerals. The sulphur beds extend for several miles, and are traceable for half a mile in width. The fissures run nearly north and south, and they seem to be connected with disturbances which have taken place at a comparatively recent period. The sulphur occurs in small globular masses in the gypseous beds, and also in a crystalline form, especially near the fissures from which the heated gas issues. These deposits have not been largely worked hitherto, and little or nothing is known about the exploration and production of the sulphur.

RUSSIA.—Although not of commercial importance, the sulphur springs at the Imperial baths of Sergiefsh, on the banks of the river Sok, a tributary of the Lower Volga, are interesting as throwing light upon the formation of the deposits we have already noticed. A considerable volume of gaseous sulphurous water issues from Permian limestones and marls, which also contain bitumen and gypsum, and forms a large pool in which the different substances are deposited with some amount of separation and distinctness.

ICELAND.—Still more interesting for the same reasons are the sulphur springs of Iceland. These are situated in a rather volcanic district near Reykjalid, in the north of the island. A series of them occur along a bare mountain-ridge covered with deep loose sand of a reddish colour, near the lake Myvatn. The sulphur springs are of two kinds, which are, however, intimately connected with each other: first the actual sulphur springs, 'solfataras,' which consist of hot gaseous exhalations from the earth, which deposit sulphur. These may be called the dry springs. The others are the mud springs or 'makkaluber.' They are boiling springs filled with muddy water of a dark colour, owing to the presence of sulphate of iron. The first-named are on the slope of the mountain; the last are at its foot, on the edge of a plain that stretches away to lava plains beyond. The first, the solfataras, occur separately and grouped in small numbers; they are characterised by a small flat elevation, from the centre of which the gaseous matter is ejected through one or more openings, which are seldom more than half an inch in diameter. The gaseous matters consist of sulphuric acid, hydrogen, sulphurous and aqueous vapour. Usually they issue quietly, and they are partially condensed in the air, and carried about in the form of a white vapour which seriously affects the respiratory organs. Occasionally the gas issues forth with great violence and with a loud hissing noise. On the surface these solfataras are covered with loose blown sand like the rest of the ground, but underneath this there is a crust of bright yellow and nearly pure sulphur, which is that taken for exportation. Underneath this, the other products of the vapours are found in pure clay, gypsum, and ferruginous clay in separate layers.

The 'makkaluber,' of which there are seven or eight large ones, have cauldron-shaped openings, eight to twelve feet in diameter; these are filled with boiling black mud, which is thrown up on the edges of the cauldron. The chemical processes that take place in these 'makkaluber' and solfataras are the same. From the former there is no deposit of sulphur on the surface, but underneath, at no great distance, there is a treacherous deposit of boiling black sulphurous clay. One of these, near the lake Myvatn, is 300 feet in circumference, and is a lake of boiling mud, with many jets scattered over its surface.

The region of sulphur springs extends in a south-westerly direction from those of ·Reykjalid and Myvatn to Krisnvik, in the south-west of the island. Here, as in the north, in the neighbourhood of the mud springs there is usually a bed from one to three feet deep of pure sulphur, resulting from the decomposition of the gases escaping from the earth on their coming into contact with the atmosphere. Mr. C. S. Forbes, in his *Iceland, her Volcanoes and Glaciers*, in describing a sulphur spring of this locality, says: 'In the valley beyond, about fifty feet beneath us, lay a huge cauldron, twelve feet in diameter, in full blast, brimming and seething with boiling blue mud that spluttered up in occasional jets five or six feet in height, diffusing clouds of vapour in every direction. If a constant calm prevailed here, instead of ever-varying gales, the sulphur sublimated from these sources would be precipitated in regular banks; as it is, it hardly ever falls twenty-four hours in the same direction, the wind blowing it hither and thither, capriciously distributing the sulphur shower in every quarter.

'Such, with little variation, save in locality, were the numerous soufrières and solfataras that we visited, and they extend over a space of twenty-five miles in extent. The riches of the district consist not so much in these numerous crusts of almost pure sulphur as in the beds of what I must be permitted to term sulphur earth, which are promiscuously scattered in all directions, ranging from six inches to three feet in thickness, and containing from 50 to 60 per cent. of pure sulphur.'

The sulphur deposits of Iceland are hardly thick or extensive enough to be commercially valuable, especially as the sulphur when obtained has to be carried a two days' journey on horseback to a trading station, where the price obtained for it is seldom more than 3*s*. per hundredweight. Hence it is that sulphur works which have been established at different times in the island have usually come to an unsuccessful end.

Springs containing sulphur, from those whose water is

taken medicinally to those more nearly approaching the mud springs in Iceland, are found in nearly all countries, and in strata of various ages. Some of these we have had occasion to refer to in describing the salt deposits.

PYRITES.

In the chapters on arsenic, cobalt, and other minerals referred to in this work, as well as in the chapters relating to copper and iron, in a former volume,[1] I have noticed in detail the composition of those mixtures of sulphur with iron and with copper, and of those with arsenic, cobalt, and antimony, which are known by the general name of pyrites. I have now to notice the mineral more particularly as a source whence a portion of the sulphur of commerce is derived, and as the chief mineral now used in the manufacture of sulphuric acid,—iron pyrites, the general composition of which is: iron 46·7, sulphur 53·3. It is, however, frequently mixed with small quantities of copper, arsenic, and minute quantities of gold.

BRITISH ISLANDS.—The production of iron pyrites in these islands in the year 1881 amounted to 43,616 tons, of the estimated value of 30,033*l.* Of this quantity, about 15,000 tons were derived from the Coal-measures, in the strata of which they are known as 'brasses.' The rest were from the older strata of Cornwall, Wales, and Ireland. The production of the copper mines of Cornwall and Devon amounted to 16,000 tons, and the bulk of these were treated for sulphur and for arsenic, in the manner to be described, at the works in those counties. Wales, from the older rocks, produced 3,679 tons, of which one-half came from the Cae Coch mine, near Trefrhiw, in Carnarvonshire, a region whose strata, slates included, are black with decomposed iron pyrites. Ireland, from four mines in counties Clare and Wicklow, from Cambrian or Silurian strata, yielded 8,598 tons, the bulk of which came from the Tigrony mine in County Clare. But the imports of pyrites from abroad far exceed the native production of the mineral, and the growth of the pyrites trade, and the treatment of the

[1] *Metalliferous Minerals and Mining.* Crosby Lockwood & Co.

mineral on the Tyne, the chief seat of the industry, are the most remarkable developments of modern times.

In the year 1810, Messrs. Doubleday and Easterby erected the first sulphuric acid chamber on the Tyne at Bell Quay. They obtained the plans of the chambers from the Messrs. Tennant of Glasgow, at which place, and at Amlwch, near the Mona and Parys copper mines, the manufacture had been carried on, to a small extent, previously. They imported the first cargo of sulphur from Sicily about the same time, and its arrival in the Tyne is described as having excited great attention. On the Sicilian sulphur there was then an excise duty of 15*l*. per ton, but this amount was remitted to the firm by the Government. In the year 1825 the duty was reduced to 10*s*. per ton, the cost of the ore delivered in the Tyne being from 6*l*. to 8*l*. per ton. In the year 1838 the King of Sicily granted a monopoly of all the sulphur raised in his dominions to Messrs. Faix & Co., of Marseilles. This proceeding led the chemists of this country to turn their attention to the use of pyrites as a substitute for sulphur. For fifteen or sixteen years the British Islands furnished all the pyrites that were required, but in the year 1856 attention began to be turned to the importation of pyrites from Belgium, Norway, Westphalia, in Germany, and from Spain. It was also found that the pyrites from Spain were the richest in sulphur, and from that time they have been increasingly used. The total importation of pyrites—iron and copper—by this country in the year 1881 was 542,378 tons, of the value of 1,202,281*l*., this quantity also being about the average quantity imported yearly for the preceding ten years. From these pyrites there were, in 1881, obtained, besides the sulphur and sulphurous products, 14,000 tons of copper, 258,463 ounces of silver, and 1,490 ounces of gold. Of the pyrites imported, 173,978 tons were received into the Tyne, 204,163 tons in the Mersey, to be used at the works grouped about Widness and Runcorn; the rest was distributed among the ports of London, Hull, Bristol, Cardiff, Swansea, Ardrossan, Glasgow, and some minor ports. We have not the particulars of quantities from the different countries exporting to us for the year 1881, but in the previous year, when the total quantity amounted to 658,047 tons, Spain sent 463,199 tons, Portugal 166,519 tons, Norway 10,952 tons, Germany 8,695 tons, and other countries 8,684 tons.

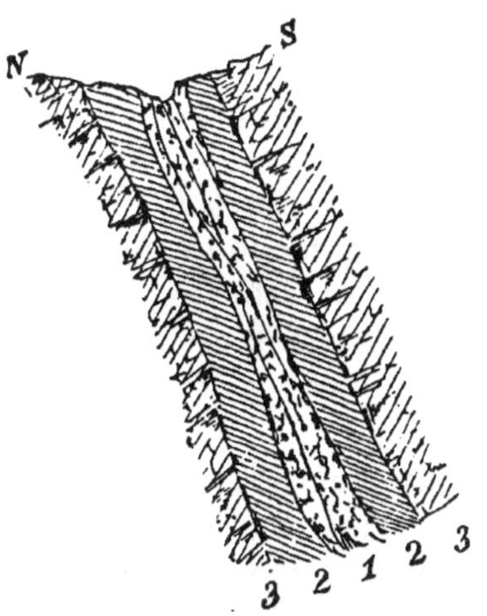

FIG. 58.—SECTION OF CUPREOUS PYRITES IN RANENFJORD, NORWAY.

1, Pyrites Deposit, 30 to 40 ft. thick. 2 2, Hornblende Schists. 3 3, Mica Schists.

The pyrites from Norway are obtained along the west coast, at the mines of Ysteroen and Vognaes. Farther north, a deposit is just now being opened near Mo, in Ranenfjord, which will serve as an illustration of the usual mode of the occurrence of great masses of pyrites in that country. Fig. 58 will illustrate this.

The German pyrites are obtained from mines to the east of the Rhine, between Cologne and Coblentz, and from the mines of the Hartz, and which are treated on the spot.

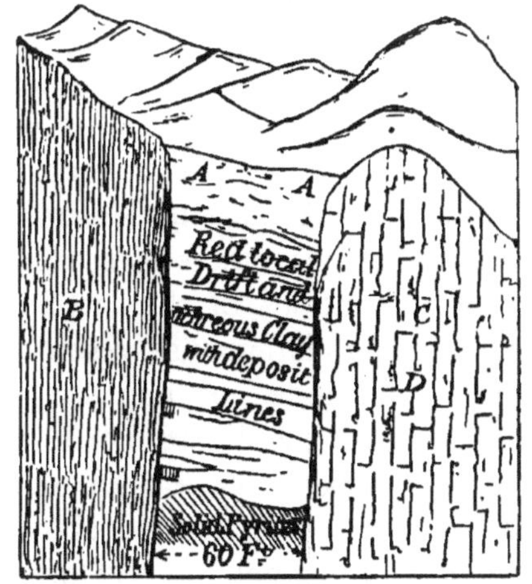

FIG. 59.—SECTION OF PYRITES DEPOSIT, HUELVA, SPAIN.

A A, Fahlband. B, Decomposed Clay Slate. C and D, Greenstone.

SPAIN AND PORTUGAL.—It will be seen that the bulk of the pyrites now used in Great Britain comes from these countries. The principal district from which the mineral is obtained is the rich mineral region situated on each side of the boundary line between the two countries, shown upon the map (fig. 72), as that line comes down to the south of the two countries, between Cape St. Vincent on the west, and the Straits of Gibraltar on the east. As described in the chapter on manganese, the rocks of the district consist of slaty schistoze strata of the Lower Silurian group. These are interstratified with greenstone and felspathic rocks of various kinds. The general direction of the strata is from north-east to south-west, with a dip of 40 to 50 degrees, and in the vicinity of the pyrites deposits up to 70 degrees, to the south-east. The pyrites occur in the line of bedding, and usually resting upon, as its lower stratum, a greenstone or felspathic rock. Fig. 59 will illustrate their usual mode of occurrence.

The pyrites deposits are not continuous, either vertically or horizontally. They occur in lenticular-shaped masses, extending in length to 450 yards, in depth 90 yards, and in breadth 125 yards, this being the present size of the great open cast worked on the south lode of the Rio Tinto mines. The proportion of sulphur in the mineral is about 48 per cent. Copper is present up to 3 per cent , and there is also a small quantity of gold and silver. The district was worked by the Phœnicians and Romans. The red colour of the river Rio Tinto indicating the presence of mineral, subsequently and at intervals the mines were worked by the Government of the country. Ten years ago, when an English company took over the Rio Tinto mines from the Government, the production was 50,000 tons a year, and the mines were worked at a loss. There are three great companies now at work in the immediate neighbourhood—the Tharsis, Mason & Barry, and the Rio Tinto—with one or two others of scarcely less importance. A glance at the operations of the last-named company will show the magnitude and immense resources of the mines. There are three chief lodes or deposits on the property—the south, San Dionisio, and the north. It is on the first of these that the workings have been most extensive. The great open cast excavation mentioned above is on this deposit, and from 500,000 to 600,000 tons of ore are annually taken from this excavation. There are underground workings on the same lode below the open cast, and from these 320,000 tons of ore were taken in 1881. These workings are reached by a tunnel driven on the lode some 500 yards in length, and this is now

FIG. 60.—GENERAL VIEW OF THE RIO TINTO PYRITES MINE, SPAIN, FROM A PHOTOGRAPH OF MODEL CONSTRUCTED BY MR. N. THWAITE, 23, COMMERCIAL ROAD, LONDON, S.E.

being connected with the San Dionisio lode, which it will open up. On reaching the principal shaft, which has been sunk on this lode, the tunnel will be two miles and a-half in length, of which a mile and a-half will be in solid mineral. Three shafts —the Albert, Edward, and Alice—have been sunk upon this lode. The operations upon the north lode are more recent, but about 700 yards of tunnel have been driven upon it, nearly 300 of which are in mineral. In uncovering the great open cast upon the south lode, nearly three and a-half million tons of earth and other over-burden were removed before the ore-deposit could be worked. The ore raised and disposed of in 1881 was 1,000,000 tons, of which 230,000 tons were shipped to England and other countries, 50,000 tons were left in store at Huelva, and the remainder was treated at the mines for the production of copper, of which 10,000 tons were produced and sent to England. There are about six miles of precipitating tanks, and 15,000 to 20,000 tons of pig-iron are annually used in effecting the precipitation of the copper. Ten thousand five hundred people are employed by the company at the mines, on the railway, and at the port of Huelva. There are forty miles of railway about the mines, which are also connected with the port of Huelva by a railway eighty-three kilometres long, which winds down the valley of the Rio Tinto. The other two companies are in very successful work, and their production and export of ore, although not so great as that of the Rio Tinto, is still very large. Fig. 60, from a photograph by Mr. Thwaite, of Commercial Road, London, S.E., and kindly given me for use in this book, will afford an idea of the extent, importance and varied nature of the works at the Rio Tinto Mines.

In the year 1882 the Tharsis mines shipped 212,218 tons of pyrites, 5,534 tons of precipitate, and 184,059 tons of iron ore. The year's operations yielded a profit of 335,676*l.*, of which 314,479*l.* was distributed in dividends, equal to 27½ per cent. on the paid-up capital.

PART IV.

METALLIC MINERALS,

ARSENIC, COBALT, MOLYBDENUM, ANTIMONY, AND MANGANESE.

(OF WHICH THE METAL IS RAPIDLY OXIDISABLE, AND HAS HITHERTO BEEN EXTRACTED WITH DIFFICULTY.)

CHAPTER XV.

ARSENIC.

Arsenic, Native—Orpiment, Realgar, Mispickel—Production of Arsenic in the British Islands—Mode of Treating the Ores in Cornwall and Devon, and in Bohemia—Mispickel of Norway and Sweden, of France, of Germany, of Austria, Transylvania, and Hungary, of Russia, Spain, Turkey, China, and of North and South America.

ARSENIC occurs native, in combination with oxygen, as white arsenic or arsenous acid, united with sulphur in its two sulphurets, *Orpiment*, or yellow sulphuret, and *Realgar*, or red sulphuret. Also, in combination with sulphur and iron, as *Mispickel*, or arsenical pyrites. It is also largely combined, as we have seen, with other minerals—lead, manganese, nickel, silver, and cobalt, as well as with lime, as arsenate of lime. Arsenic burns with a bluish flame, and in all the forms in which it is found it may be unmistakably known by the white fumes and the pungent garlic smell it gives off when heated.

NATIVE ARSENIC is found in massive, granular columns, and in rhombohedral crystals. In colour it is tin white, light lead grey, and tarnishing to greyish black. Its specific gravity is from 5·65 to 5·75, but when artificially prepared the gravity is described as greater. It is of a very brittle and friable nature, and is easily pulverised. When freshly fractured it has a brilliant appearance, but it soon oxidises on exposure to the air. The fracture is granulated, and sometimes a little foliated and splintery. It rises in vapour at 356°, without undergoing fusion, and burns with a pale bluish flame when heated just below redness. It occurs in connection with silver and lead ores.

White Arsenic (*Arsenous Acid*).—This is the same substance as the arsenic sold in the shops, and it is a well-known poison. Chemical composition: arsenic 75·8, oxygen 24·2. Specific gravity 3·7. It is white in colour, astringent, and sweetish in taste, and soluble in water.

Sulphuret of Arsenic occurs in two forms:—

1. *Orpiment*, or *Yellow Sulphuret of Arsenic.*—This occurs massive, also disseminated and crystallised in oblique four-sided prisms. The crystals are small, with a smooth surface, and are irregularly clustered together. Colour, fine golden yellow, with brilliant pearly or metallic lustre on the face of the cleavage. Specific gravity 3·3 to 3·5. Chemical composition: arsenic 61·0, sulphur 39·0, but different tests have shown a range of from 57·0 arsenic, and sulphur 43·0, to arsenic 84·0 and sulphur 16.

2. *Realgar*, or *Red Sulphuret of Arsenic.*—Also occurs in oblique four-sided prisms, and, but more rarely, in a massive form. The colour ranges light aurora red, scarlet red, and orange yellow, with a resinous, translucent, or transparent lustre. Specific gravity 3·35 to 3·65. Chemical composition: arsenic 70, sulphur 30, but these proportions are varied by small quantities of iron and silica. Both of these sulphurets burn with a blue flame on charcoal, and evaporate entirely before the blow-pipe with a strong garlic odour. The foregoing ores are soft, being easily scratched with the finger nail.

Mispickel (*Arsenical Iron Pyrites*).—Occurs massive and in long rhombic prisms, silver-white colour, with a grey or yellowish tarnish, and a black streak. Chemical composition: arsenic 46·0, sulphur 19·6, and iron 34·4. These proportions are varied by small quantities of silver and gold, and with cobalt up to 9 per cent. This mineral forms the source whence a large quantity of the arsenic of commerce is derived.

Arsenic is largely used in the arts and manufactures. Besides its use as a poison, white arsenic is used as a flux in the making of glass, and also to give a porcelain-like appearance to the same article. The sulphurets are extensively used for colouring purposes, and attention has often been drawn to the danger, real or supposed, arising from their use in the colouring of wall-paper. The colour known as King's yellow is made from *orpiment*, and an ammoniacal solution of the same is used for dyeing. *Realgar* is used in the manufacture of fireworks, more especially for the production of the flame known as white Indian fire. A combination of *white arsenic* with oxide of copper produces the fine green colour known as Scheele's green. It is also used in the proportion of 1 per cent. with lead in the manufacture of shot. Its presence causes the molten lead to separate more easily into drops as it is poured down through a sieve from the top of a lead-tower. The chief repositories of the ores of arsenic are the following:—

BRITISH ISLANDS.—The production of refined arsenic in the counties of Cornwall and Devon has been on an extensive scale for nearly a century. In the year 1880 the production was 5,738 tons, of the value of 43,498*l.*, and in 1881 the yield was 6,156 tons, of the value of 45,070*l.*, being an average value of about 7*l.* 5*s.* 6*d.* per ton. The finer kinds realise 10*l.* 10*s.* per ton. This last quantity was produced by fourteen mines, worked otherwise chiefly for copper. Of these twelve were in Cornwall and two in Devon. In Cornwall, the largest quantity was produced at the Greenhill works, 1,628 tons, and in Devonshire, the Devon Great Consols Copper mine produced 2,851 tons. The general composition of the ore whence this arsenic is derived is, arsenic 42·88, iron 30·04, and sulphur 21·08; but these quantities are varied by proportions of copper, and sometimes tin. In obtaining the arsenic from the ore, the latter is first calcined in revolving calciners, usually of a large size, 30 feet long and 5 feet diameter. These consist of iron tubes lined with firebrick. They are inclined about 7 in. to 1 ft. 6 in. from the horizontal, and are kept revolving—where practicable, with water-power. The ore is put in at the upper end, and gradually passes downwards, parting with its arsenic by the way. These calciners are said to require but little fuel, the heat being largely kept up by the combustion of the arsenic itself. The crude arsenic is collected in long flues. Sometimes it is sold as so collected, containing

80 to 90 per cent. of arsenic. More frequently it is resublimated four times in reverberatory furnaces and collected as absolutely pure *arsenous acid*, white as snow, in flues and chambers. It is then ground in a mill to a fine powder, which finds its way down flexible pipes to a hole in the barrels waiting to receive it. These barrels are subjected while filling to a shaking motion, which causes the arsenic to settle down compactly, and the flexible pipes are so closely fitted to the holes in the barrels that no dust escapes, and in some works there is no smell whatever of arsenical fumes. Each barrel contains about 3½ cwt. Much the same process is practised at Joachimsthal, in Bohemia. The ores are roasted, the arsenic caught in long flues and chimneys, and sublimated a second time, usually with a little potash. Probably there is less attention to some points of detail here, for it is said that the process is very destructive to human life, few of the persons employed in the manufacture living beyond the age of thirty to thirty-five years.

NORWAY AND SWEDEN.—In Norway arsenical pyrites mixed with cobalt up to 35 per cent., occurs at Skuterud. An extensive bed of mispickel also occurs near Kongsberg, which contains a little gold. The mineral also occurs as arsenic silver in the great silver mine at Kongsberg. In Sweden I have seen vast beds of mispickel from 3 to 6 feet thick, extending over the surface of the ground in the region between the towns of Ludvika and Fahlun. Similar deposits occur in the country between the towns of Norköping and Nyköping, on the coast of the Baltic, and we shall see how largely the mineral is associated with the cobaltiferous ores of Gladhammar, farther south on the same coast.

In FRANCE native arsenic has been worked at the mine of St. Marie, probably now belonging to Germany. It occurred uniformly over a considerable distance between the hanging side of the vein consisting of slaty rock and the calcareous matrix of the lode itself.

In GERMANY the sulphuret and oxide are found to some extent in the lead mines of the Hartz district, but all the ores are more abundant in the mines of silver and tin worked on the Saxon side of the Erzgebirge, near Freyberg. Native arsenic exists in rounded masses composed of concentric layers, and containing up to 4 per cent. of silver. The sulphurets also occur in the vein stuff of the different lodes, and associated with the other ores worked. The same is true of arsenical pyrites.

On the AUSTRIAN side of the mountains, at Joachimsthal, the oxide of arsenic occasionally appears in the form of quadrangular prisms, and, more rarely, as a thin white efflorescence. As I have already intimated, the manufacture of arsenic from arsenical pyrites is carried on here.

In TRANSYLVANIA *realgar* has been found as a vein a foot thick, traversing a dolomitic limestone. Near Nagyag it occurs in irregular masses, accompanying gold and silver ores. Near Felsobanya realgar occurs in the form of rectangular prisms.

In all these mining districts orpiment also occurs, sometimes in the form of small foliaceous masses. At Tayoba, in Hungary, it is found in small detached masses, as confusedly grouped together eight-sided crystals, and imbedded in a bluish clay.

In RUSSIA native arsenic occurs in the Siberian mines in large masses.

In SPAIN oxide of arsenic is found associated with cobalt in the valley of Gistain, in the Pyrenees.

In ITALY realgar occurs as a volcanic production. It is found near Naples in the form of eight-sided crystals in the fissures of volcanic rock. It occurs also in the lava ejected from Vesuvius in the year 1794, and is found under similar conditions near Guadaloupe, in the West Indies, where it is known as red sulphur.

Both realgar and orpiment are found in Kurdistan, in TURKEY in Asia, the latter being also obtained from CHINA.

In the UNITED STATES OF AMERICA native arsenic occurs at Haver Hill, N.H., in mica slate, and also at Jackson in the same state.

The sulphurets are found in SOUTH AMERICA.

CHAPTER XVI.

COBALT.

Cobalt, Origin of Name—Description of its various Ores—Commercial Varieties—Cobalt in the British Islands—Cornwall—North Wales—The Foel Hiraeddog Mine—Norway, Skuterud Mine—Indications about Kongsberg and Drammen—Sweden—Mining District, from Nyköping to Westervik—The Cobalt Mines of Tunaberg, of Gladhammar—Processes employed at the latter Mine for the Extraction of Cobalt from the Ores—Mines of Hvena—Germany, Riegelsdorf, Annaberg and Schneeberg—Austria, Joachimsthal—Past Production of Cobalt in Bohemia—Spain—Mine in the Pyrenees—France—America—Imports into England—Suggestions.

THE name of this mineral is derived from *Kobbold*, which was the name given in Germany and throughout Scandinavia to the evil spirits of the underground regions in general, and the evil genius of miners in particular. It got this evil repute both from its deceptive appearance, once looking so much better than before its uses were discovered it was proved to be, and more than this, spoiling the copper ores with which it was associated. The ores of this mineral were of little or no use until the middle of the sixteenth century, when they were first employed to impart a blue colour to glass, and subsequently for the colouring of porcelain and earthenware. For this purpose they are still very valuable. It is said that one grain of cobalt gives a full blue colour to two hundred and forty grains of glass. For these purposes the mineral is best adapted in a finely powdered state, the oxides of the metal being most suitable for use.

In its metallic state, in which it is not found in nature, it is of a pale steel grey or tin colour, with a bright lustre. Its more compact ores also have a metallic appearance, with

a specific gravity ranging from 6·2 to 7·2. The looser ores are of a reddish colour, and have a specific gravity of about 3.

Until recently it has not been used in its metallic state, but probably it will become of great importance in this form, in consequence of the discovery within the last year or two by Dr. Fleitmann, of Iserlohn, Germany, of its value in the art of plating other metals, and for which in its metallic form it is now used, but in a limited extent, on account of its high price. In appearance cobalt plating is superior to that of nickel. The ores of cobalt are as follows:—

SMALTINE (*Tin White Cobalt*).—Chemical composition: cobalt 18 to 25 per cent., arsenic 69 to 79 per cent. In colour tin white or steel grey. Its varieties are *Cobaltine*, containing 33 to 37 of cobalt, with varying proportions of sulphur and arsenic. This variety has a silver white colour with a reddish tinge.

RADIATED WHITE COBALT, or *Chloanthite*, containing 9 to 14 per cent. of cobalt, and cobalt pyrites, or sulphuret of cobalt, or Linnæite.

EARTHY COBALT (*Black Oxide of Cobalt, Asbolane*).—Chemical composition: usually oxide of cobalt 24, oxide of manganese 76, of a bluish black to black colour, in form both earthy and massive.

COBALT BLOOM (*Erythrine, Arseniate of Cobalt*).—Chemical composition: 37·6 of oxide of cobalt, with 38·4 of arsenic acid, and water 24·0. Of a peach and crimson red colour, varying to greenish grey; possesses a foliated structure like mica. It is also called peach-blossom ore, and red cobalt ore. This is usually found in thin layers, or in small cavities and in aggregated crystals. Its varieties are *Roselite*, of a rose red colour; *Arsenite of Cobalt*, a compound of oxide of cobalt and arsenous acid, resulting usually from the decomposition of other cobalt ores; *Sulphate of Cobalt*, or cobalt vitriol, consisting of oxide of cobalt, sulphuric acid, and water.

It will be seen that in nature this mineral is largely associated with arsenic, but its ores are distinguished from arsenical pyrites by the blue colour they give with borax under the blowpipe.

The commercial varieties of cobalt are known as Zaffre, Smalt, and Azure. Zaffre is prepared by calcining the ores in a reverberatory furnace, by which the sulphur and arsenic are driven off, and the oxide of cobalt remaining is mixed with twice its weight of powdered silica. Smalt and azure, which have a rich blue colour, are made by fusing zaffre with potash and glass. Perhaps the best form in which the mineral is now sent into the market is that produced by the chemical process now in use at the Gladhammar mines in Sweden, as described farther on. I will now describe the principal deposits of cobalt ore in the world, including some that were formerly worked, for the sake of comparison as to their mineralogical conditions.

THE BRITISH ISLANDS. *Scotland.*—A considerable quantity of cobalt bloom was, towards the close of the last century, obtained from the copper lode worked for both copper and silver at Aloa, near Stirling, and traces of the mineral have been observed in the débris of the lead and copper mine of Newton Stuart, in Galloway.

Cornwall.[1]—In the year 1754 the Society for the Encouragement of Arts and Useful Discoveries awarded a premium of 30*l.* for the best cobalt found in England to Mr. Beauchamp, who mined some in Gwennap. The lode in which this was found also contained bismuth, which was thrown away until Dr. Albert Schlosser, who visited the mine in 1775, separated the cobalt from the bismuth and preserved both. The cobalt was discovered in a branch of a lode, while driving an adit upon Pengreep Estate, but it did not hold in depth. Cobalt had at that time been found at Wheal Trugo, near St. Columb Major, in a vein 4 to 6 inches thick, where it crossed a copper lode, but the cobalt only continued a little way from the point of intersection. This ore was considered worth 60*l.* a ton. Discoveries of small quantities were also made at Dudman's mine, in Illogan, at a mine near Ponsnooth, and in Dolcoath, which in a later period has produced good ore. Subsequently the mineral

[1] See De la Bêche, *Geological Report on Cornwall and Devon.*

has also been found near Botallack, at Polgooth, and St. Austell, as well as in a cross-course and adjoining a copper lode near Redruth. Unfortunately the mineral has not been discovered in sufficient quantities to yield much profit.

North Wales.—The only cobalt mine at present worked in Great Britain, is the Foel Hiraeddog mine, near Rhyl, Flintshire.[1] The ore is found in one of the numerous cavities that occur in the lower massive limestones of the Carboniferous series, and which are locally known as "swallows." These "swallows" or "pockets" contain in this neighbourhood deposits of iron ore (hæmatite), and it was in working this pocket for that mineral that Mr. Gage, the proprietor, discovered the presence of both cobalt and manganese in the ore. He had noticed some black strings in the limestone, and on testing them with the blow-pipe, he found that the black colour was due to oxide of manganese in some cases and to oxide of cobalt in others. This led to a further examination of the ore of the pocket, which led to the discovery of the presence of cobalt ore of the nature shown in the following analysis :—

Samples.	1.	2.	3.
Cobalt sesquioxide	37·40	20·63	26·20
Nickel sesquioxide	8·58	6·85	10·35
Manganese binoxide	23·12	39·50	25·58
Iron sesquioxide	13·85	15·10	21·10
Copper oxide	traces	traces	0·25
Silica	0·45	2·00	0·60
Alumina	0·10	0·50	0·18
Water	16·00	15·00	15·00
	99·50	99·58	99·26

The nature of the deposit will be explained by a reference to fig. 61. 1 is the irregular cavity or crack widened out in the limestone, stretching downwards to a depth of 240 feet or more, and varying in width from a string to 8 or 10 feet. It has

[1] Le Neve Foster, B.A., D.Sc., F.S.S., 'On the Occurrence of Cobalt Ore in Flintshire,' *Cornwall Geological Society*, 1880.

horizontally a north-north-east and south-south-west direction, and an extent of about 30 yards. For the most part it is perpendicular in depth, but varies occasionally to the east and to the west. 2, 2, is red clay, which contains lumps of hæmatite ore, and lumps and grains of wad, earthy manganese, and asbolane, or earthy cobalt ore. The lumps are sometimes as large as a walnut or hen's egg, and when broken show a reniform or botryoidal structure; they are of a black colour, and are soft enough to mark paper, and give a shining streak upon porcelain. 3, 3, are loose fragments of limestone imbedded in the clay, and 4, 4, are the limestone beds themselves. The whole of the clay is not cobaltiferous, the whole of the width of 10 feet being sometimes without any cobalt. The presence of cobalt in the clay is detected by taking a piece of the latter and drawing a portion of it over a piece of porcelain with the flat side of the blade of a knife, when, if a series of black shining streaks are formed, it is concluded that cobalt is present. Even this is not a certain test, as the earthy ore of manganese will produce similar streaks, so that chemical tests have frequently to be resorted to. The deposit has been followed down by a series of small shafts, and the only preparation of the ore at the surface is the picking out of it the lumps of iron ore and fragments of limestone.

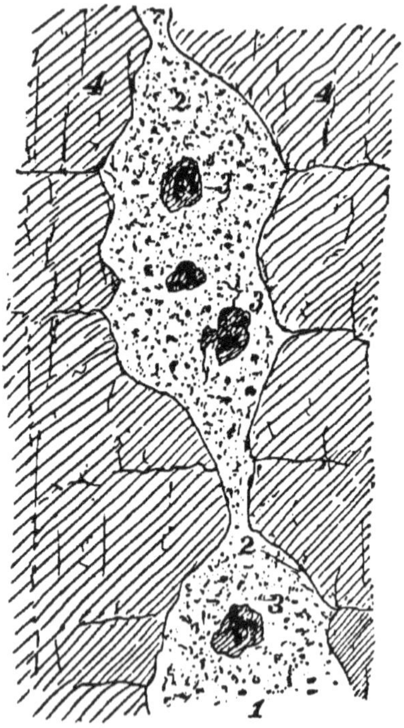

FIG. 61.—SECTION OF CAVITY IN CARBONIFEROUS LIMESTONE, FLINTSHIRE, CONTAINING COBALTIFEROUS IRON ORE.

In this state its percentage of cobalt is very small, only ranging from 1·0 to 1·08 per cent., and the nickel from 0·4 to 1·1 per cent. The production and value of the ores have been as follows in the years given :—

Year.	Tons.	Cwt.	Value. £	s.	d.
1878	98	18	 616	17	0
1879	116	11	 833	2	5
1880	49	3	 297	6	4
1881	63	14	 309	12	8
	328	6	2,056	18	5

This gives the average value of the ore at 6*l.* 5*s.* per ton. It will be seen that the value was greatest in the first year. With regard to the origin of the ore, iron is largely present in the reddish beds of the limestone, and still more so in beds of red shale and clay that form a marked feature in the limestone ridges all along their course. There are also some extensive deposits of iron pyrites, one very marked one occurring at the end of the Tarlagoch lead mines, at the end nearest Foel Hiraeddog. In the chapter on manganese I notice how largely manganese is disseminated throughout these limestone beds. We have not, therefore, far to look for the origin of the ores of the pocket. The hæmatite and the manganese are probably derived from the limestone beds and red shales. The cobalt may either have been originally associated with the manganese, or it may have been derived, which seems more probable, from the pyrites deposits, especially as a lump of the latter, that has been found at a depth of 30 ft. in the mine, has shown upon analysis traces of cobalt and copper.

The discovery of cobalt and nickel in this pocket leads to the inference that similar pockets, of which there are many, may be mineralogically similar, and hence deserve a careful examination.

NORWAY.—Crossing the North Sea we find that cobalt has been, more or less, worked for the last hundred and ten years in Norway. A discovery of the mineral was made on the estate of Skuterud, in the parish of Modum, about fourteen English miles from Drammen, in the year 1772. The ore occurred in beds (fahlbands), interstratified in gneissic strata like other mineral deposits of this age. The chief ore was cobaltine, which often occurred in a crystalline form. There were also arsenical pyrites containing at times as much as 10 per cent. of cobalt. There were in the same beds various ores of copper and other minerals. These all occur in gneissic rock. On the discovery of the ore the estate was purchased by the king, and works were established upon it in the year 1783, under German management. From 1827 to 1840 the mines were carried on by a private firm with considerable success, until the introduction of artificial ultramarine for a time paralysed the industry. An English firm purchased the works in 1849, when the stock of cobalt sold for 8*s*. 6*d*. per lb. At the present time the mines and works belong to a Saxon company, and in the year 1882 the production of calcined cobalt ore amounted to 160,000 lbs.

Similar beds, with copper pyrites, iron pyrites, arsenical pyrites with cobalt ores, occur in the gneissic rocks around Kongsberg, of which perhaps the most important are those until recently worked, and the ores smelted near Hougsund, between the towns of Kongsberg and Drammen.

SWEDEN.—Passing eastward into Sweden, we find deposits of cobalt ores associated with the copper ores, which in beds and lodes abundantly occur in the highly mineralised district that extends for some miles inland on the shores of the Baltic, from Nyköping on the north to below the town of Westervik on the south.

In the north of this district there are what for many years were the important mines of Tunaberg. These mines are situated about twelve English miles south of Nyköping, and about two miles from the sea on the bay of Vik, at the head of which is the town of Norköping. The rocks are, as usual, gneiss, through which a vein or lode runs in an east and west direction. The lode is largely filled with limestone; possibly it may be a limestone bed in the gneiss, in which copper pyrites is but sparsely sprinkled, as is also arsenical cobalt pyrites or cobalt bloom, but not abundantly. Galena is also present in small quantities. In the gangue of the vein or bed there is also a beautiful variety of green-coloured felspar, varying from light to dark shades of that colour, and aggregated in clusters of crystals. Serpentine also occurs, but rarely.

An analysis of the cobalt ore obtained gave the following result :—

Cobalt	36·66
Arsenic	49·00
Iron	5·66
Sulphur	6·50
Loss	2·18
	100·00

The lode contracts and widens, and the quantity of ore it contains is very variable. Owing to this fact, the mine has passed through many stages. In the latter half of the last century it was vigorously worked with water-wheel pumps and other requisite machinery. During the present century it has been intermittently worked with varying success. In sorting the ores the lumps and crystals of cobalt are picked out, carefully ground to powder, and packed in bags for the market, chiefly England; and it will be remembered that this mine differs from the next to be described by the separate occurrence of the cobalt from the copper ores. It is possible, however, that with increased knowledge it will be found that cobalt is more or less present in the copper pyrites themselves.

At the southern end of the district referred to, and about ten English miles south of the town of Westervik, are the Gladhammar copper and cobalt mines, from which at the present time large quantities of cobalt are derived.

These mines, a fine model of whose workings is to be seen in the museum at Stockholm, were opened as far back as the fifteenth century. They were worked first for iron, and at a later period on several occasions for copper, but they were always abandoned on account of the cobalt and nickel mixed up with the ores, the uses of these metals being then unknown; they were the Kobbolds, the evil spirits of the mine. Interesting relics of these early workings are seen in the tunnels and chambers which were driven and opened by means of fire, before powder came into use for mining purposes. I had a very pleasant visit to these mines in the summer of 1880, and perhaps I cannot do better than transcribe the description of the mines given to me on that occasion by the chief engineer and chemist, Herr Alfred Hasselbom, of Göteborg, through whose energy and chemical skill the mines had been brought into successful working for cobalt.

'The mines are some distance from the Fårhult Station, on the Westervik and Hultsfred Railway. The metalliferous beds and deposits have a very great extent, stretching from the Ryss mine in the north-west to the neighbourhood of Lund, in the south-east, the whole length along which the principal ores are found being about 8,000 feet.

'The rock constituting this field is a quartzy eurit, that is to say, a rock containing quartz as its essential part, and which kind of rock is much renowned in Scandinavia as carrying deposits of pyrites, copper, nickel, and iron. The mineral deposits have a strike or direction from the north-west to the south-east, and have an inclination of 10 degrees from the perpendicular to the south-west, the rock itself having the same direction and inclination.

'The rock constituting the matrix in which the different ores are embedded is chiefly chlorite, commonly intermixed with hornblende and with magnetic iron ores. Mica also often occurs as matrix, which is especially the case in the Odelmark mine. In some cases the minerals have the quartz rock itself for their matrix.

'The deposits of ores occur as beds and layers of varying extent and width, and generally the different beds are connected with each other through smaller cross veins, so as to form a complete network of ores stretching throughout the entire field; but some of these connecting veins are too thin to be of any commercial value. The presence of magnetic pyrites in all these veins accompanying the ores of cobalt and nickel is well displayed by the inclinations and variations of the com pass.

'The principal minerals occurring in the Gladhammar mines are the following, viz. :—

'Cobaltine in a pure state, with 30 per cent. of cobalt bloom, erythrine, and arseniate of cobalt. The other ores found are copper pyrites, occurring in all the beds, and rarely metallic copper, iron pyrites—always carrying cobalt; magnetic iron pyrites, zinc blende—very scarce; galena is found chiefly in the Holtandare mine; hæmatic iron ore, which with magnetic iron ore is so constantly mixed with cobalt that it may be considered the chief matrix of the cobalt ores. In the north-west part of the field in the Ryss mine antimonium ore is found accompanying the copper pyrites. In the southern part of the property molybdenum glanz is found on the débris heaps left by the ancient workers.

'The principal mines worked, together with their peculiar mineral features, are as follows :—

'*Bonde Mine*, which is about 70 ft. deep. Cobalt and nickel pyrites are found here, having a thickness of 3·5 feet, and they are intermixed with some zinc ores and galena. The continuation of this mine to the north-west is the

'*Holtandare* or *Baggen Mine*, about 120 feet deep, yielding cobaltine, cobalt pyrites, with copper and iron pyrites. All of these ores are very rich and abundant. There is also galena and iron ore as the usual matrix for cobalt.

'*The Svensk Mine.*—This mine has been actively worked for a long time, and it has yielded a steady supply of cobalt and nickel, with copper pyrites and iron pyrites containing nearly 2 per cent. of cobalt. This is an extensive deposit that continues undiminished. It also seems to get richer in depth. It is to this mine that an adit level some 600 feet long was driven in ancient times by means of fire, probably to work the copper deposits. Running parallel to this deposit, about 20 feet to the east, is a valuable deposit of rich cobalt and copper. This is worked as the

'*Odelmark Mine.*—Tin white cobalt with cobalt pyrites,

giving an average of 15 per cent. of cobalt, are obtained from this bed, and the yield is very large. This mine is worked to a depth of 100 yards. North-west of the Svensk mine, and on the run of the same bed, is the

'*Knut Mine*, which has been continuously worked for some time. It produces the same ores as those of the Svensk.

'*Ryss Mines*.—These are about 1,500 feet north-west of the Knut Mine. They are old mines, and they have recently been re-opened. They yield copper pyrites and iron pyrites with cobalt, besides antimonium ore, *Boulangerite*.'

Some of the ores from the various mines are beautiful examples both of single and combined ores.

Since the mines have been worked for cobalt, large quantities of both cobalt and copper have been obtained by screening the old waste heaps, and it is computed that there still remains in these heaps about 50,000 pounds of metallic cobalt to be extracted by the process lately adopted.

Until within the last few years the ores were simply picked or screened, and sold in this state, and the mines were successfully worked on this plan. Three years ago, however, machinery and appliances were erected, under the direction of Mr. Hasselbon, for treating the mixed ores chemically, and which is now in successful working. The process is generally as follows:—

The ore when brought out of the mine is picked, and is broken to about the size of apples by a Blake's stone crusher. It is then calcined by mixing it with slack, and slowly burning it at a low heat until the sulphur and arsenic are driven off, or partly so. Then it is ground very fine in a mortar-mill, after which it is placed in a row of furnaces with an addition of alkaline matter. It is placed, first, in the furnace furthest from the fire, and is gradually moved forward to the hottest place. After it is taken out of the furnaces and cooled it is lifted to the top of a building in which there are three tiers of round tanks about 9 feet in diameter and 8 feet deep. There are ten of these tanks in each story of the building, and the fine-burnt ore is first placed in those of the uppermost tier, in which, as in all these, is water kept hot by steam being driven into it.

The iron is precipitated in the uppermost series of tanks, the liquid, with the contained copper and cobalt, being run off into the middle series, in which, by means of the addition of scrap-iron, the copper is precipitated. The liquid with the contained cobalt is run into the lowest series of tanks, and from there it is pumped into square compressors with flexible sides, and with fine gauze partitions; a pressure of 100 pounds to the square inch is put on, during which the cobalt is caught and pressed into flat cakes between the gauze divisions, through which the water flows off clear. The cobalt now appears as a fine compact yellow substance, the cakes of which are taken from the press separately. The cobalt is next burnt and becomes black oxide, when it is fit for the most important purposes of manufacture. The copper is taken out of its tanks or vats, and, separated from fragments of iron, becomes a precipitate with from 80 to 90 per cent. of metallic copper. The iron is calcined and forms Indian Red. The refuse is carried through a series of twelve wooden tanks 8 feet long, 4 feet wide, and 4 feet deep, the result being ochres of various degrees of purity. These are ground upon the spot, and are ready for the painter. About 130 persons were employed at the mines at the time of my visit. The ochres and paints have a large sale in Sweden, Denmark, and Norway. The cobalt goes for the most part to Saxony.

There are also cobalt mines at Vena, or Hvena, near Orebro, in Nerika. These were started in 1809, and for some time their exact position was not known. In 1880 the production of cobalt ore from these mines was 70,000 lbs.

GERMANY.—An interesting series of deposits of cobalt occur, and have been worked more or less for more than a century, at Riegelsdorf, in Hessia, Germany. A series of veins or faults cut through the limestone and other beds that overlie the copper slate bed, which is famed not only for its copper, but also for the impressions of fish which abound in it, and around which the copper ore is richest. As these cracks come down upon the copper slate bed, they are charged with cobalt, which dies out upwards.

On the Saxon side of the Erzgebirge, near Annaberg and

Schneeberg, cobalt ores are obtained from mines worked for other minerals.

AUSTRIA.—Cobalt has been obtained in considerable quantities from the mines of Joachimsthal, on the Bohemian side of the range of mountains just named. Here it is met with combined with the silver ores, and also in separate masses. As in Sweden, the ores of cobalt were here thrown away as useless, but since their value has been discovered the débris heaps have been carefully picked over. About a hundred years ago the yearly production of cobalt in Bohemia was given at 1,000,000 lbs.; at the present time it is not likely to be so much.

In SPAIN, a cobalt mine has been worked in the valley of Gistain, in the Pyrenees. A gneissic rock is interstratified with beds of silicious and micaceous schists. Over this there is a bed of red felspar, on which rests a bed of dark bituminous schist of a friable nature, and varying in thickness from 30 to 60 feet. This bed is traversed by veins of cobalt which run from east to west, and range in thickness from half an inch to five feet; near the surface the ore is earthy cobalt, and in depth, arseniate of cobalt. The sides of the veins are also penetrated with cobalt. The veins do not pass out of the schist into either the red felspar or the limestone.

FRANCE.—On the French side of the Pyrenees cobalt was discovered in a vein of quartz that traversed a mass of ferruginous shale. The mine yielded large quantities, and the manufacture of the ores was conducted in works erected upon the spot.

The mineral has also been worked in the Vosges, where it occurs in veins, having for its matrix crystallised carbonate of lime. It has also been found associated with silver ores in the mines of Allemont, in Dauphiné.

AMERICA.—The mineral occurs plentifully in the State of Missouri, where it has been mistaken for black oxide of copper. It is largely obtained from the mine La Motte, associated with manganese. The ores are exported to England and refined there. The ores of the mineral are also obtained from Carolina. An analysis of samples of these showed oxide of cobalt 24, oxide of manganese 76.

The mineral is also found in smaller quantities in various other places.

Unfortunately no return has been obtained for some years past of the imports of cobalt and several other minerals into England, but I am indebted to the kindness of Mr. Robert Hunt for the following information.

In the years given below the imports were as stated:—

	1859.	1863.	1867.	1870.
	Tons.	Tons.	Tons.	Tons.
Cobalt	16	1	5	44
Ditto, ore	486	446	427	10
Ditto, oxide	2	16	28	31

In the year 1879, among the ores unenumerated, were from—

	Tons.
Norway	973
Germany	215

These returns include cobalt and nickel; in all probability about 400 tons of cobalt.

The selling price of black oxide of cobalt in this country is from 10*s*. 6*d*. to 12*s*. 6*d*. per lb.; so that, with the increasing uses there are for the metal, any further source from which it could be obtained would be a boon to manufacturers.

I would suggest that in this country the clay pockets in the Carboniferous limestone be well examined. In Sweden the débris from the numerous copper mines that have been worked in the district I have described would probably, if tested, be found to contain considerable stores of the mineral.

CHAPTER XVII.

MOLYBDENUM—ANTIMONY.

Molybdenum, description of its Ores—Commercial Uses—British Islands: Inverness-shire, Charnwood Forest, Calbeck Fell—Norway: Arendal, Numedal—Sweden—Description of the Deposits of Ekholmen, on the Baltic Coast—Germany—Austria—Hungary—America—Antimony, Early Knowledge and Uses of—Story of the Origin of its Present Name—Native Antimony—Ores of Antimony—Uses of the Mineral—Antimony in the British Islands—Cornwall—Sweden: Sala Mine, Ofverrud Mine, Gladhammar Mines—Germany: Hartz and Erzgebirge—Austro-Hungary—Borneo—Algeria—America—New South Wales.

MOLYBDENUM.

It can hardly be said that Molybdenum has as yet been completely reduced to a metallic state. When most nearly approaching to this condition it is of a steel grey colour, with a specific gravity of 6 to 6·5. It rapidly oxidises on exposure to the air, especially with heat.

Molybdenum is not a widely disseminated mineral, and hitherto it has been obtained in very limited quantities as compared with other minerals. It occurs in nature in three forms—first and chiefly, as a sulphuret; more rarely, secondly, as an oxide; and thirdly, and more rarely still, in association with lead.

1. Molybdenite (*Sulphuret of Molybdenum*).—Of a pure lead grey colour, with greenish grey streak. Specific gravity 4·5 to 4·75; occurs in lumps and masses, and in six-sided prismatic crystals, also in thin foliated plates and laminæ. Chemical composition: molybdenum 59·0, sulphur 41·0; infusible before the blow-pipe, but gives off sulphur fumes; par-

tially soluble in nitric acid, leaving a small residuum. It has the appearance of plumbago, and has a soft, greasy feel.

2. MOLYBDIC or MOLYBDENA OCHRE, of an orange yellow or sulphur colour; contains molybdenum up to 8 per cent.

3. MOLYBDATE OF LEAD (*Yellow Lead Ore*).—Chemical composition: protoxide of lead 59 to 63; molybdic acid 30 to 35, with small proportions of silica, and occasionally oxide of iron; colour, straw or honey yellow, with a waxy or resinous appearance. Occurs in masses, and also crystallises in four-sided tables, and also in eight-sided prisms.

For the quantities of the mineral raised, the ores of molybdenum have in different combinations a varied use. In the manufacture of pottery molybdenum blue or blue carmine is used to impart a blue colour of great brilliancy and durability. This preparation is obtained by mixing molybdate of sodium with a solution of chloride of tin, when a blue precipitate is obtained, which, when dried, forms the colour referred to, and is ready for use. With tin salt as leys, this molybdenum blue can be used on wool and silk. A solution of molybdic acid on sulphuric acid is also used for dyeing silk a brilliant blue. Molybdate of ammonia is used for various chemical purposes, and in Sweden molybdate of sodium is used in medicine for the treatment of dropsy.

BRITISH ISLANDS.—In some of the Cornish copper and tin mines this mineral is occasionally met with, but scarcely in commercially paying quantities. Many years ago it was worked in the older rocks of the western part of Inverness-shire. It occurred in chloritic schists. I have seen it covering the smooth joint faces of the syenitic or fine-grained granite rocks of the Mount Sorrel quarries of Charnwood Forest. It has also been raised at Calbeck Fell, in Cumberland.

NORWAY.—In this country it is found associated with copper ores in the neighbourhood of Arendal on the south coast, and in those of the long valley of Numedal, leading up from Kongsberg to the north-west. In both these localities the rock is a hornblendic gneiss. In SWEDEN important deposits have recently been opened and worked in the rich mineral district extending along the coast of the Baltic, elsewhere referred to. I am able to describe these deposits more minutely.

They occur on the island of Ekholmen (Oak Island), a little island situated in the Archipelago of Westervik, on the south-eastern part of Sweden, and about twenty miles north of the city of Westervik. The area of the island is 1,500,000 square feet. The strata of the island consist of hornblendic gneiss and micaceous rocks. Through these strata, from north-west to south-east, and dipping to about 70 degrees to the south-west, run seven distinct veins. These veins or lodes vary in width from 6 inches to 2 feet. At one point four of these veins coalesce and form one deposit 5 feet wide. The contents of these veins are molybdenite, molybdic ochre, and copper pyrites, with a gangue of felspar and quartz. The lodes have a known length of 270 feet, and, where worked, they have been proved to a depth of 30 feet. Molybdenite, in a quite pure state, occurs in lumps weighing up to 5 lbs. Where it occurs in smaller fragments and particles it is screened without difficulty to a state of great purity. In the summer of 1880, from the 2nd of June to the 2nd of October, three men raised from these lodes 1,400 lbs. of pure molybdenite, together with about 10,000 lbs. of second molybdenum ore, having an average of 9 per cent. molybdenum, with about 5,000 lbs. of unscreened ore. There is an absence in the contents of these lodes of phosphorus, wolfram, and other minerals which it is difficult to get rid of. The ore raised was sent to Germany, which is the chief market for nearly all the most valuable ores of Norway and Sweden. The prime ore realised 16*s*. per kilogramme (very nearly 2 lbs.), or about £25 per cwt.

In GERMANY the mineral is found to a limited extent in the mines on the Saxon side of the Erzgebirge, imbedded in quartz and in a hard greenish marl.

On the AUSTRIAN side of the same mountain range it is found at Zinnwald and Schlackenwald, in Bohemia. It occurs in quartz, which at the first place named is greasy and opaque, and at the latter place it occurs in plates in transparent quartz.

In some of the mines of HUNGARY it is found in small rounded masses, like those of Ekholmen, which are deposited in a greyish coloured clay—very likely decomposed from the surrounding rock. These masses are composed of large shining plates, closely adhering to each other. They contain a proportion of silver ranging up to 12 per cent. of the mass. In other mines in Upper Hungary molybdenum is found associated with gold.

In the UNITED STATES of America it occurs at Haddam and Saybrook, in Connecticut; at Blue Hall Bay, in Maine; at Shutesbury and Burnfield, in Massachusetts; near the Franklin Furnace, New Jersey; and near Warwick, New York.

ANTIMONY.

ANTIMONY is a mineral which has been known from remote times. It was the *Stibium* of some of the ancients, and was much valued as a dye for personal use. The sulphuret of antimony was also known as *Alcofal*, an Arabic word for a very fine powder, in which condition it was used for the adornment of the face. The words alcophal, alcosol, and alqufocor, the name given to the fine powder of the sulphate of lead used by the potters, as well as alcohol, are probably derived from the same source. The more modern name, antimony, is said to have arisen from the experiments of Basil Valentine, a German monk. Basil having tried the effect of the mineral upon the pigs, found that after a preliminary violent purging they grew fat upon it. He therefore assumed that his brother monks would thrive upon a similar treatment. The dose he gave them unfortunately produced a very different effect, and they all died. The medicine then received the name of anti-monk, whence it passed to antimony. The mineral occurs in nature in the following forms :—

NATIVE ANTIMONY.—In its metallic state, in which, however, it occurs but rarely, antimony is of a brilliant tin or silvery-white colour, with a slight tinge of blue. It is usually associated with a little silver or iron; it is crystalline in structure, and is very brittle, and possesses a highly lamellated structure. It fuses readily before the blow-pipe, and at a temperature a little above that of zinc. If the heat is increased it boils and passes off in fumes. Its specific gravity is from 6·6 to 6·75. The native metal soon loses its lustre on exposure. It may be produced from its sulphuret by mixing 4 parts of the latter with 3 parts of crude tartar and 1½ parts of nitre, and placing the mixture in small quantities in a red hot crucible. Antimony is closely associated with several other metals—cobalt, arsenic, copper, iron, zinc, silver, and lead. The presence of the metal is generally supposed to deteriorate the metals it is associated with. Thus, in the language of the miners, it "robs" the lead with which it is found. The native metal was first discovered in the silver lead mines of Sala, Westmannland, Sweden.

SULPHURET OF ANTIMONY, STIBINE, ANTIMONITE, GREY ANTIMONY.—Colour, lead grey with a grey streak and a blackish shining tarnish. Chemical composition: antimony 71 to 73, sulphur 27 to 29. Occurs in various forms, massive and granular; also in thin laminæ, and crystallised into fibrous and radiating groups; brittle in texture, but, in a thin laminated form, slightly flexible; fuses in the flame of a candle, and vaporises rapidly before the blow-pipe. This is the chief ore from which the metal with its preparations is derived, and it comprises the following varieties. As we may not have occasion to refer to some of them again the localities where they are found are given.

1. *Arsenical Antimony.*—Colour, tin white with a reddish streak. Chemical composition: antimony 36·4, arsenic 33·6; occurs massive and granular. From Allemont and Bohemia.

2. *Berthierite, Hardingerite.*—Contains 9·8 to 16 iron, 52 to 62 antimony, and 29 to 31 sulphur; colour, dark steel grey with a yellowish or reddish tinge. From Auvergne and Anglas in France, Braunsdorf in Saxony, Tintagel and Padstow in Cornwall.

3. *Boulangerite.*—Colour, dark lead grey, inclining to blue, with a dark streak of a silky metallic lustre, finely granular, and in fibrous, radiating and columnar masses. Chemical com-

position: 24 to 26 antimony, 18 to 19 sulphur, and 56 to 58 lead. From Molière, in France, Gladhammar mines, Sweden, Ober Lahr, Lapland. Also found in Siberia.

4. *Feather Ore, Plumosite.*—Colour, dark lead grey. Chemical composition: antimony 31, lead 50, sulphur 19. From the eastern part of the Hartz.

5. *Geokronite, Kilbrickenite.*—Colour, pale lead grey with a slight tarnish. Chemical composition: lead 67, with 1· to 2· of copper and iron, antimony 16, arsenic 4·7, and sulphur 17. From the Sala mines, Sweden, Mérédo, in Spain, and near Pietrosanto, Italy. *Kilbrickenite*, from county Clare, Ireland.

6. *Jamesonite.*—Colour, steel grey to dark grey; occurs in parallel or radiating prismatic crystals. Chemical composition: antimony 36, lead 44, sulphur 20. From Cornwall, Estramadura, in Spain, Hungary, Siberia, and Brazil.

7. *Kobellite.*—Colour, blackish lead grey to steel grey, with a blackish streak; structure, columnar and radiate. Chemical composition: sulphuret of bismuth 33, sulphuret of lead 46, and sulphuret of antimony 13. Hvena, in Nerik, Sweden.

8. *Plagionite.*—Colour, dark lead grey; occurs in oblique rhombic crystals, also in thick tabular forms. Chemical composition: antimony 37 to 38, lead 42 to 43, and sulphur 21. From Wolfsberg, in the Hartz.

9. *Steinmannite.*—Colour, lead grey; occurs in cubes with cubic cleavage, and massive. Chemical composition: varying proportions of lead, sulphur, and antimony, with a little silver.

10. *Zinkenite.*—Colour, steel grey with a bluish tarnish, in six-sided needle-like prisms; also fibrous and massive. Chemical composition: antimony 44, lead 35, sulphur 21. From Wolfsberg, in the Hartz.

The whole of these ores are soft, are easily scratched with the finger-nail. They have a specific gravity of 5·4 to 6·6.

WHITE ANTIMONY, VALENTINITE.—Colour, yellowish grey, greyish white; also brown, grey, and peach blossom red, with a white streak with a pearly or adamantine lustre. Chemical composition: 83·6 antimony, and 16·4 oxygen; occurs in rectangular crystals and in long tabular masses. Becomes yellow in the flame of a candle, and fuses to a white mass; soft, like the preceding ores; gravity, 5·57. It has varieties—

1. *Antimonate of Lead.*—A rare mineral of a yellow colour, ranging from grey and green to black. Chemical composition: oxide of lead 61·8, antimonic acid 31·7, water 6·5. Occurs near Nertschinsk, in Russia.

2. *Red Antimony.*—Composed of both sulphuret and oxide of antimony. From the Hartz, Saxony, and Hungary.

3. *Romeine.*—Antimonate of lime, of a honey yellow colour, and hard enough to scratch glass. Occurs in groups of minute eight-sided crystals. Found in Piedmont.

4. *Senarmontite.*—Colour, white to grey, more or less transparent, brilliant, resinous lustre. Composition similar to that of white antimony, but differs slightly in crystallisation. From Algeria.

Antimony has been used from very early times for purposes of supposed personal adornment, being more particularly used by eastern ladies for darkening the eyebrows. It is also said to have been one of the first minerals used in medicinal preparations. Its use was proscribed in France in the year 1566 on the ground that it was poisonous, and in 1609 a physician was expelled the faculty for having administered it. The prohibition was withdrawn in 1650, the mineral having then recently been received into the number of purgatives. In 1668 a provision was made by which the use of it was limited to doctors of the faculty. It is now generally admitted that although some of its preparations are virulent as emetics, it may, with intelligence and care, be used safely with advantage. The tartar emetic of the apothecary is a mixture of antimony and potassa. As a paint for the bottoms of ships its oxide is very valuable, the use of it for this purpose being limited by its price. Among the more modern uses to which the mineral is applied is that of forming alloys with other metals. It hardens and improves the quality of tin, and is used in the manufacture of Britannia metal and pewter wares generally, Britannia metal being usually composed of 100 parts of block-tin, 8 parts of antimony, and 2½ parts of copper, or 2 parts of copper and bismuth. An alloy of 17 to 20 per cent. of antimony with lead makes the most approved type-metal, the larger the type the smaller the proportion of antimony. Small proportions of bismuth and copper are sometimes added. It is used with great success in the composition of alloys adapted to withstand great friction, as machinery blocks and bearings, and also for those used in the manufacture of scientific instruments, as well as in the manufacture of sheathing metal for ships, and of shot, shell, bullets, and balls.

In the British Islands antimony has been profitably obtained from the mines of North Cornwall, and more particularly from the neighbourhood of Endellyon. Borlase refers to it as being obtained in that parish in 1758, and also those of St. Stephen and St. Austell. In the three years from 1774 to 1776 inclusive, according to Price, 120 tons of antimony were raised in Cornwall, of which 95 tons were raised at Wheal Boys Mine in Endellyon, the price being from 13*l.* to 14*l.* 14*s.* per ton. The remainder was obtained from a mine near Saltash or Tredinnick, which has also subsequently been worked for the mineral. From 1800 to 1840 workings were carried on at intervals near Endellyon, at Trevatham, near St. Teath and St. Merryn.[1] The metal was discovered in Pillaton about the year 1819, 20 tons being raised in that year, 33 tons in 1820, and 79 in 1881. The principal ore was Jamesonite. In the year 1778 there were antimony works at Restronget Creek, Falmouth Estuary, the ore being the ordinary sulphuret. About the year 1856 Mr. James Bennet made a valuable discovery of the ore on Lady Molesworth's land, in St. Keev's parish. A recent discovery of the mineral near Liskeard shows, according to an analysis by Mr. M. W. Bawden, 60 to 70 per cent. of antimony, with 6 to 13 ounces of silver, to the ton of ore. The mineral occurs in the ordinary clay slate of the country in which the lead mines are worked, and it occurs in true lodes in the usual way. There have not been any returns of the mineral as raised during the last two years.

SWEDEN.—Both native antimony and sulphuret occur, the

[1] *Geological Report on Cornwall and Devon.* By Sir H. De la Bêche.

former but rarely, in the Sala Mine, in the north-west of Sweden, Westmannland. They occur in association with lead ores, very rich in silver, that are found in lodes which traverse a primitive limestone from east to west. The mineral is also mixed with the grey copper ore of the Ofverrud Mine, in the parish of Glafva, Wermland. The occurrence of the mineral as in the form of boulangerite, along with cobaltiferous copper ores and ores of lead at the Gladharnmar Mines, near Westervik, is referred to in the chapter upon cobalt. Some of the specimens of the associated ores are very beautiful.

GERMANY.—Antimony ores occur in the two great mining districts of this empire, the Hartz and the Erzgebirge, in nearly all the varieties described. It is associated with the lead ore of the former district, and with the more mixed ores of the latter. It is also produced to some extent as sulphuret near Schemnitz, Kremnitz, and Felsbany, in the Austro-Hungarian Empire. Some years since the production of antimony in Germany reached 73,500 lbs., and in Austro-Hungary 539,000 lbs.

BORNEO.—This island has for some years past been the chief source of the supply of the mineral. Little is as yet known of the mineralogical conditions in which it occurs, except that it consists chiefly of antimony ochre, of which large quantities are shipped to England, to Hamburg in Germany, and Boston in the United States of America, where it is refined; also to China. The chief mining districts are in Sarawak. Native antimony in a very pure state is also obtained. The exports of the mineral in the year 1881 from this State amounted in value to $72,516.

ALGERIA.—In this country, at Ani-bebbouch, in Constantine, sulphuret of antimony, as senarmontite, is largely worked, and is shipped in considerable quantities to England. It is found as octohedral crystals and in fibrous masses, and as botryoidal incrustations of a snow-white colour.

In AMERICA antimony has been found in small quantities in the United States at Carmel, Maine, Cornish and Lyme, New Hampshire, and at Soldier's Delight, in Maryland. In the province of New Brunswick mining operations have been conducted at Prince William. The ores occur in a number of parallel contact veins or deposits, occurring between different strata. These run east and west, and they have a number of small feeders running north and south. The ore is sulphuret, with native antimony rarely occurring.

In NEW SOUTH WALES there were in the year 1881, 862 acres of land let for the mining of the ores of antimony, the value of the ores obtained during the previous ten years being 11,830*l*.

The lodes on the Munga Creek, near the Macleay River, traverse Devonian strata and contain the oxide and sulphide of antimony. The gangue of the lodes consists chiefly of quartz, and in this the ores of antimony occur in irregular bunches. Associated with the ores at Armidals there is free gold which is visible to the naked eye. The yield of antimony from various parts of Australia has largely increased during the last few years, and the mining of the mineral is becoming an important feature in the industry of the country.

CHAPTER XVIII.

MANGANESE.

Manganese in its Native State—Rapid Oxidisation of—Uses of the Ores of Manganese—Alloys of the Metal with Iron and Copper—Description of its Ores—Manganese Ores of Great Britain and Ireland—Cambrian Rocks of North Wales, of Scotland—Silurian Strata of Ireland, of North Wales—Silurian and Devonian Strata of Cornwall and Devon—History of Manganese Mining in those Counties—Devonian Strata of North Wales—Carboniferous Limestone of Derbyshire, of Shropshire—The Manganese Deposits of Nassau, North Germany—Italy: Mines of Val d'Aosta and Tournanche—Spain, Mines of the Huelva District, of Cape de Gata—France: the Romanèche Mines, Cevennes, Vosges—Occurrence of Manganese in the Mines d'Asprières — America — Manganese Ores of Missouri—Arizona—Canada—Inferences.

In its metallic state manganese is a greyish white metal of considerable brilliancy and of a granular texture, with much the same appearance as hard cast-iron, and of a very brittle texture. It is, however, an operation of much difficulty to extract the metal from its ores. Hydrogen and charcoal at a red heat reduce the superior oxides of this metal to the state of protoxide without eliminating the pure metal at that temperature, but at a white heat charcoal deprives the metal of the whole of its oxygen.

Manganese oxidises very rapidly on exposure to the atmosphere, falling down in a black powder, and, as will be seen, it is in various stages of oxidisation that the mineral is chiefly found in its ores. The black oxide of manganese was for a long time known as *magnesia nigra*, from a fancied resemblance to *magnesia alba*. Its true nature was first made out by Scheele in 1774, and almost immediately afterwards Gahn obtained from it the metal now known as manganese.

The ores of manganese are used in the arts for the generation of oxygen and the manufacture of bleaching powder. The sulphide and chloride of manganese are used for colouring purposes in the printing of calico, the sulphate imparting a chocolate or bronze colour. Its ores are also used in glass manufacture, chiefly for giving a violet colour. Latterly the application of these ores has been considerably increased and extended by their use in the manufacture of various valuable alloys in conjunction with other more purely metallic minerals. Iron, for example, readily unites with manganese at a high temperature, and a proportion of the latter mineral makes the iron whiter and harder. It is also found that iron ore containing a small proportion of manganese is the best for the manufacture of steel. Nor is the tenacity of the iron destroyed by the admixture of a small portion of manganese, as much as 1·85 of the metal having been found in a bar of iron of good quality; and traces of the metal are found in good iron and steel from Russia, Sweden, and France. At the smelting-works at Dillenburg, in Hesse-Nassau, several valuable alloys of the metal are made with iron, copper, and tin. Mansfield refined copper, for example, mixed with 11 per cent. of manganese, forms the pure manganese bronze, which is capable of bearing a heavy breaking strain. A mixture of copper 85, tin 6, zinc 3, and cupro-manganese 3 parts, gives a casting that will bend to a right angle before showing fine cracks. An alloy also of great hardness, but workable with tools, is also made at the same works with 80 parts of copper, 10 parts of tin, and 10 parts of manganese. Varying proportions of the mineral are used with iron, tin, copper, and zinc to produce results adapted to particular uses. The presence of a little iron in manganese imparts to it magnetic properties, and renders it less liable to rapid oxidisation, but the presence of manganese in any force destroys the magnetic properties of iron.

The ores of manganese are the following, their specific gravity ranging from 3·4 to 5·2.

PYROLUSITE, from the Greek *pur* fire, and *luo* to wash, in reference to its use in taking away the green and brown tints of glass. Colour, iron black with a black streak. Chemical composition: 63·6 manganese, and 36·4 oxygen. This is the most abundant ore of manganese. It includes the varieties *Varvarcite* and *Polianite*, which are of the same chemical composition, but differing—the first by containing a little water, and the second by its less hardness. Crystallises in small rectangular prisms. Specific gravity 4·8 to 5·0.

PSILOMELANE, Greek *psilos*, smooth or naked, and *melas* black. Colour, greenish black, bluish black, and black, shining reddish or brownish black streak. Chemical composition rather varied, 4·7 to 11 protoxide of manganese, 50 to 80 hyperoxide of manganese, 6 to 16 baryta, 2 to 5 potash, 0 to 1 copper, and 0·5 protoxide of cobalt. Associated in the same mines with pyrolusite, the two often occurring in alternate layers. It is an abundant ore. Its varieties are *Heteroclin* and *Marcelline*, or *Braunite*, containing 10 to 16 per cent. of silica.

MANGANITE.—Chemical composition: 89·9 of peroxide of manganese and 10·1 water. Occurs in rhombic prisms, and also in a massive form. Specific gravity 4·3 to 4·4. Colour, dark steel grey to iron black, often brownish black, with a tarnished brown streak.

MANGANESE SPAR (*Rhodonite*).—Occurs in oblique rhombic prisms and also in large masses. Colour, dark rose red, bluish red, or reddish brown. Specific gravity 3·5 to 3·6, translucent, vitreous, or pearly. Not affected by acids. Chemical composition: 45·33 silica, and 53·67 manganese protoxide, with 3 to 5 lime, and 0 to 6 iron protoxide. Its varieties are *Bustanite*, from Mexico, of a pale greenish or reddish grey colour, with 14 lime; *Fowlerite*, from New Jersey, with 7 to 11 iron protoxide; *Tephroite*, also from New Jersey, with 29·8 of silica, and 70·2 of protoxide of manganese. Less definable varieties are, *Paisbergite* from Sweden, and *Hydropite, Photicite, Allagite*, and *Horn Manganese.*

CUPREOUS MANGANESE.—Colour, black, inclining to blue and black. Specific gravity 3·1 to 3·2. Opaque, rather brittle or friable. Chemical composition: 14· to 17· copper protoxide, 1·6 baryta, 2·5 lime, 0·2 to 0·6 protoxide of cobalt and nickel, 15 to 17 of water, with frequent admixtures of other substances.

CREDNERITE.—Chemical composition: 42·85 copper protoxide, and 57·15 peroxide of manganese. Colour, iron black with a brownish black streak.

TRIPLITE (*Ferruginous Phosphate of Manganese*).—Occurs in a massive form. Colour, blackish brown. Chemical composition: protoxide of manganese 33·2, protoxide of iron 33·6, phosphoric acid 33·2, with a little lime. Its varities are *Heterosite*, with 41·77 per cent. of phosphoric acid, and *Huraulite*, with 38 of phosphoric acid, and 18 per cent. of water.

WAD (*Bog Manganese*).—Earthy or compact; also occurs in coatings and dendritic forms. Colour and streak brown or black. Chemical composition very varied, but consists usually of from 30 to 70 per cent. of peroxide of manganese, mixed with varying proportions of peroxide of iron, and the oxides of copper and cobalt, and with from 20 to 25 per cent. of water. Like bog iron ore it is formed in marshy places from the decomposition of substances containing manganese.

Besides the above there are *Hausmannite*, a sesquioxide of manganese, containing when in a pure state 72 per cent. of manganese, found in Thuringia and Alsatia. *Peloconite* from Chili, a mixture of manganese and iron. *Manganblende* or *Alabandine;* chemical composition: 31 protoxide, and 69 peroxide of manganese, or 72·4 manganese, and 27·6 oxygen, the proportions varied by proportions of sulphur, occurs with gold in Transylvania. *Hauerite* from Kalinka, near Neusohl, in Hungary; colour, reddish brown to black; chemical composition: 46·28 manganese, and 53·72 sulphur. *Diallogite* or red manganese; chemical composition: 62 protoxide of manganese, and 38 carbonic acid, varied by the carbonate of lime 0 to 13, magnesia 0 to 7, or iron 0 to 15; the Hartz, Saxony, and Hungary. Its variety, *Wiserite*, is silky and fibrous and contains water. There is also the arseniuret of manganese, greyish white in colour with a black tarnish; chemical composition: 42·75 manganese, and 57·25 arsenic, with a little iron, found in Saxony.

GREAT BRITAIN AND IRELAND.—In the British Islands we find these ores in a widely disseminated form, and also in masses sufficiently concentrated for successful mining. In the former state it permeates the Cambrian slates of North Wales, or rather the metamorphic rocks by which they are traversed, forming the beautiful moss and tree-like forms sold to visitors to the principal slate quarries as 'landscape stones.' It is gathered into small nests, bunches, and strings in the porphyritic and ash beds of the Llandeilo strata of the Principality. The lower dolomitic beds of the Carboniferous rocks of England and Wales are also covered and permeated with the same dendritic or tree-like forms, and the ores are gathered into cracks and cavities. There are also curious pockets of loose sand with a manganese nucleus and radiations in the thick sandstone beds of the millstone grit, and a reference to the chapters in this work descriptive of the phosphate of lime deposits of the world will show how persistently and universally this mineral has been associated with the ordinary qualities of phosphate of lime through all time. Let us now notice it as it is gathered into deposits sufficiently large to induce attempts at mining, beginning with the oldest strata in which such attempts have been made.

Cambrian Rocks.—About thirty-five years ago there was considerable activity in Carnarvonshire in seeking for manganese ores in the strata associated with the slate rocks of that county, without, however, leaving any permanently practical results. The deposits occurred in strings and small nests here and there, but not extensive enough for working.

In Scotland, in rocks of a similar or, it may be, older age than Aberdeenshire, a vein of manganese of considerable extent was discovered and worked for some time in the latter part of the last century. The ore was, however, extremely hard, being largely mixed with quartz and baryta, and it was further intimately combined with silica, and altogether it was extremely difficult to reduce it to powder. It is on record that this mine yielded some beautiful specimens of the radiated gray ore.

Silurian Strata of the South of Ireland.—The Glandore Mine, near Cork, has been worked for some years in slaty rocks. The ore lies in what appears to be the junction of two lodes. It is mixed with spar and some iron ore. The production of this mine in 1881 was 250 tons, of the value of 435*l.*

Lower or Cambro-Silurian Rocks of North Wales.—Mining operations have been carried on at irregular intervals in the slaty rocks of the region between Festiniog, Trawsfynydd, Bala, and Dolgelly, with their interstratified ash, felspathic and porphyritic rocks. In the slaty rocks the ores of manganese occur in irregular veins or joints, which open and close at intervals along their course, and at the intersection of these with floors or horizontal joints in the rocks. Hitherto the cost of carriage in that district has been fatal to the successful mining of so low priced a mineral, and it is very doubtful if the ore exists in sufficient quantities and in masses concentrated enough for profitable mining. Probably now that the region is opened up by railways new attempts will be made.

Silurian and Devonian Strata of Cornwall and Devon.[1] —Borlase refers to manganese as raised on Tregoss Moors in the year 1754, and mentions it was used for glass making and the manufacture of Egyptian ware in Staffordshire, for which purposes it seems to have been wholly employed until it came into use for bleaching. About the years 1760–70 manganese ores were found at Upton Pyne, near Exeter. Other deposits were discovered about the same time at Newton St. Cyres, in the same neighbourhood. This group of mines for some years yielded from 200 to 300 tons a year, at prices ranging from 30*s.* to 3*l.* per ton, but they became exhausted, Upton Pyne about 1810, and Newton St. Cyres soon afterwards. But about the same time similar ores were discovered and worked at Doddescombleigh, Ashton, Criston, and other places. About the year 1815 manganese deposits were discovered and worked in the neighbourhood of

[1] De la Bêche, *Geological Report of Cornwall and Devon*, p. 609.

Tavistock and Launceston. Subsequently the ore was raised in the vicinity of St. Stephens, and discoveries have continued to be made in the original localities and to the north-east and south-west of them until now.

Between the years 1804 to 1810 the quantity of manganese shipped from Exeter amounted to 3,000 tons a year. In 1821 the quantity raised in the two counties was estimated at about 4,000 tons. In 1839 the production was taken at 5,000 tons, of the very high value of 8*l.* per ton. At the present time there are sixteen manganese mines in the two counties, five of which produced, in 1881, 1,855 tons of ore, of the declared value of 4,922*l.*, or rather less than 2*l.* 14*s.* per ton. Of these mines, Chillaton and Hogston, Ruthen, Lydcock, and West End Down, Whetstone and Rose Exbridge, the first-named, produced 1,224 tons. The average price is, however, reduced by the inclusion of 140 tons of manganiferous iron ore at Rose Exbridge which only realised 13*s.* 9*d.* per ton. The ore from Chillaton and Hogston realised 3*l.* 10*s.* per ton.

Concerning the geological age of the manganese mines of Devon and Cornwall, it may be stated broadly that the mines extending from Launceston and Tavistock to the south-west occur in Silurian strata, and those to the north-east in beds of Devonian age. There is a considerable difference in the modes of the occurrence of the ores in these two groups of rocks. In the older or Silurian strata the ores occur chiefly as in rocks of the same age already described in North Wales, in veins, chiefly cross crosses from lodes of other minerals, irregular fissures, and at junctions of the same with lines of bedding. In the Devonian strata the ores occur chiefly in irregular masses that occupy portions of the strata themselves, and are probably of contemporaneous age. Where the ore is collected into previously worn cavities and fissures, there are often beautiful examples of crystallisations.

Several of the ores of manganese are found throughout the two counties—braunite near Launceston, psilomelane at Upton Pyne, and bisilicate of manganese near Tavistock and Callington—but the prevailing ore is pyrolusite, or grey and black ore, containing 70 to 79 of peroxide of manganese. This is combined with variable quantities of silicious and aluminous matter and oxide of iron.

The ores are associated with those of iron, but as in most other mines, the two classes of ores are distinct, and occupy separate parts of the same deposit, the ores of the Rose Exbridge mine forming an exception.

The Devonian Strata of North Wales.—At the base of the Carboniferous limestone of Denbighshire, along its course from the Eglwyseg cliffs near Llangollen to near its termination on the mainland in the promontory of the Great Ormes Head, there are nearly continuous patches of Devonian strata, corresponding probably to the upper division of that group. These strata consist of dark red sandstone, conglomerates of pebbles derived from the older rocks, hard shales, and impure limestones. The whole series rests upon a waterworn floor of the Wenlock beds of the Upper Silurian strata.

Deposits of hæmatite have at various times, including the

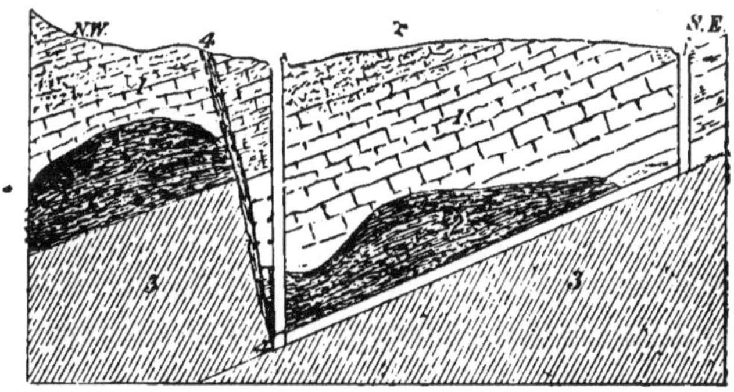

FIG. 62.—SKETCH SECTION OF NANT UCHAF MANGANESE AND HÆMATITE MINE, NEAR ABERGELE, NORTH WALES.

1, Beds of Conglomerate, Red and Blue Marl, and impure Limestone, part of the Devonian Series. 2, Manganese and Hæmatite Deposit. 3, Wenlock Shale, Silurian. 4, Fault.

present, been worked in these beds. Within the last few years a rather extensive deposit of manganese ore, associated with hæmatite, has been discovered at Nant Uchaf, near Abergele. From this deposit, in 1881, 305 tons of manganese ore were raised, of the average value of 1*l.* per ton.

Fig. 62 will give an idea of the nature of this deposit. It shows the hæmatite and manganese ores as occurring in irregular masses in the impure limestone beds. These masses owe their origin either to an original deposition contemporaneous with the disposition of the beds themselves, or by the subsequent infiltration of mineral matter into cavities made in the limestone. From their extent and position, as well as from the light thrown upon these more ancient deposits by the newer ones of Nassau and elsewhere, to be described, I incline, in the absence of evidence to the contrary, to the first supposition. This deposit is interesting from its stratigraphical relation and similarity in other respects to the Devonshire mines, showing, as it does, similar mineralogical conditions in strata of the same age in widely-separated areas.

Carboniferous Limestone, Derbyshire.—Manganese has been mined for a long time in the Carboniferous limestone of Derbyshire and in its immediately overlying clays of local derivation. It was formerly supposed to be an iron ore, but its true nature has been known for over a century. It occurs chiefly as wad, in earthy masses not unlike balls of soot, crumbling to pieces on exposure to the atmosphere, but it shows when broken a few traces of its metallic nature in the shape of numerous minute shining veins. The deposition of the manganese in bulk is of an age subsequent to that of the deposition of the limestone. It seems to have been washed out of the latter, through which it was originally distributed, occurring more abundantly near the lines of bedding, and to have been deposited in cracks, irregular fissures, and worn surfaces of the limestone beds. In the year 1881 eight Derbyshire mines yielded an aggregate of 474 tons of ore, of the value of 710*l.*

Shropshire.—A manganese mine was worked some years ago at Pant, near Oswestry, which is interesting as illustrating the Derbyshire deposits just referred to, and as throwing some light upon the Nassau deposits, to be described. A waterworn hollow in the rather impure limestone was filled with a compact white clay of pre-glacial age. At the time of my visit a short tunnel had been driven and a chamber opened in the clay. In excavating this chamber several nests and pockets of man-

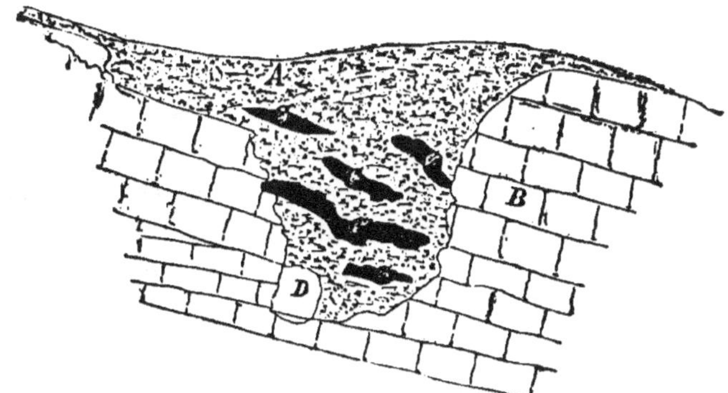

FIG. 63.—SECTION OF A MANGANESE MINE AT PANT, NEAR OSWESTRY, SALOP.
A, Stiff clay. B, Carboniferous Limestone. C C, Pockets of Manganese. D, Entrance of Level.

ganese were found, and others remained in the sides of the chambers as shown in fig. 63.

I have already observed that these limestone rocks in the neighbourhood abound with dendritic impregnations of manganese, and that the millstone grit beds are dotted with pockets containing loose sand and a manganese nucleus. The manganese of the Pant Mine, therefore, seems to be the washing out of the mineral from the rocks to which it originally belonged, an operation that was continued through a long period of time. We also see how the manganese was accumulated into distinct masses in the clay, partly as the result of chemical affinity and partly of mechanical action.

The total production of manganese in the British Islands in the year 1881 was 2,884 tons, of the value of 6,441*l.* There was also imported from other countries during the same period 18,743 tons, of the value of 71,149*l.*, or nearly 4*l.* per ton. The discrepancy between the value of British and foreign ore may be partly accounted for because with the British we have given its value at the mine, and in the case of the foreign its value when brought to port. It is also unlikely that low-class ores would be shipped at all from foreign mines.

We will now proceed to notice the manganese deposits of other countries, whence these importations are derived.

The various ores of this mineral are found in the older rocks of Sweden, of Russia, and of those of the chief mining districts of Austria and Germany, under conditions similar to those in which they are found in the British Islands. In Russia, a valuable deposit of manganite is worked at Tagel, in the Urals, and the mineral is used in the manufacture of steel. Figs. 64 and 65 give a plan and section of the manganese mine worked in the older slaty strata at Elbingerode, in the Hartz mining district, Germany, which was long successfully worked. The figures will explain the nature of the deposit, and I will pass on to notice the more recent, comparatively speaking, deposits of the Lahn Valley, in Nassau, North Germany.

GERMANY.[1]—The deposits of manganese just referred to occur in the valley of the river Lahn, a river that enters the Rhine about three miles above Coblentz. They are generally co-extensive with the deposits of phosphate of lime described in this work. They stretch from the village of Baldwinstein, between the towns of Nassau and Diez, north-eastward by Limburg, and as far east as the town of Hadamar, by Dehren, Diet Kirchen, and round by Heckolshausen to the town of Weilburg, and on to Wetzlar and Giessen. The basement rocks of the whole district, as shown in the section, Fig. 36, are shales, slaty rocks, and sandstones belonging to the Lower and Middle Devonian strata. These are surmounted by a dolomitic limestone, No. 5 of figs. 67, 68, 69, 70, 71. This is surmounted in limited areas by shales (*Cypridinen Schiefer*) of Upper Devonian age. Over the greater part of the whole area there is a deposit of compact clay of from 10 to 50 feet in thickness (No. 3 of figures). This in its turn is overlaid by a drift composed chiefly of quartz detritus (*Quartz Geschiebe*) No. 2, and *Dammerde*, or soil No. 1. Immediately upon the limestone there sometimes rests a sand-bed (No. 6), which seems to consist of the sandy particles originally belonging to

[1] F. Odernheimer, *Das Berg und Hüttenwesen im Herzogthum*. Nassau. The author also examined several of these deposits in the summer of 1867.

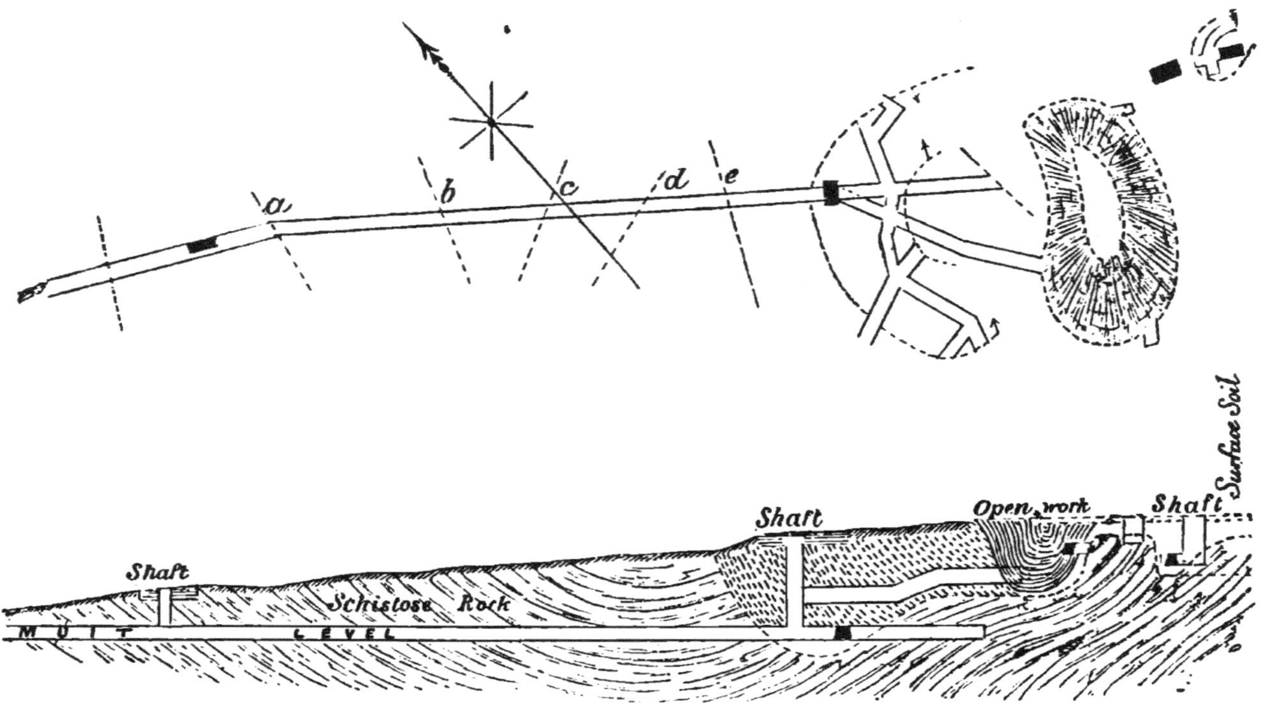

FIGS. 64 and 65.—PLAN AND SECTION OF MANGANESE MINE NEAR ELBINGERODE, GERMANY.
a b c d e, Lines showing Strike of Rock on Surface.

the limestone, the calcareous matter having been dissolved and carried away.

It is upon the worn edges of the limestone, as shown in figs. 67 to 71, that the deposits of manganese, accompanied by those of hæmatite as well as those of apatite, as described in the chapters relating to that mineral, occur. The limestone itself is frequently impregnated with manganese. It also contains little cavities in which are crystals of carbonate of lime, manganese spar, and pyrolusite of a dark blue or grey colour. The ores of manganese contained in the layers and nests in the clay above the limestone consist of pyrolusite, psilomelane, manganite, and wad. In the layers pyrolusite and psilomelane are in separate nests; wad and manganite occur promiscuously. Ironstone of several varieties, but chiefly hæmatite, are found associated with the ores of manganese. As shown in the accompanying figures, they generally occur in distinct layers, but ores of iron are not unfrequently found adhering to those of manganese. Where there are two layers of manganese the lowest is usually more mixed with ironstone than the uppermost, besides which the ore of the top bed is better in quality.

The figures 67 to 71 will illustrate the various modes of the occurrence of the ores. Fig. 66 is the plan of a hæmatite deposit near Niedertiefenbach, and figs. 67 and 68 show respectively a longitudinal and cross-section along the lines intersecting the plan. It will be seen that there are here large deposits of both manganese and hæmatite. It is also interesting to notice how distinct the two deposits are, and that where the hæmatite is at its greatest strength the manganese fails, and that the hæmatite fails where the manganese is present in full force.

In fig. 71 we have a section of the Höchst Mine in the same locality. Here there is interposed between the limestone and the manganese deposits a considerable thickness of shale, known to the miners as schaalstein, but which must not be confounded with the schaalstein containing cyprina. There are here two manganese layers, and in the schaalstein there is a continuous layer of brown ironstone.

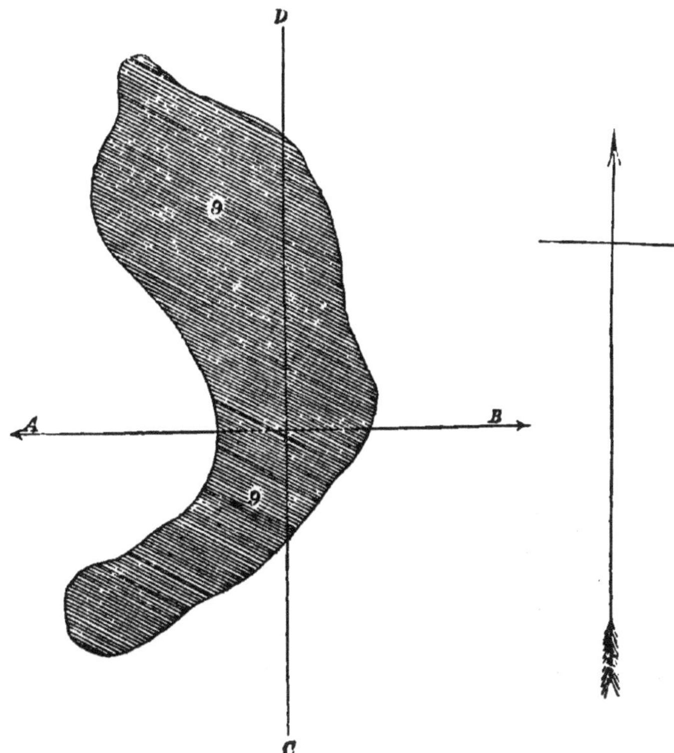

FIG. 66.—PLAN OF HÆMATITE DEPOSIT IN STEETERWASEN MINE, NEAR NEIDER-TIEFENBACH.

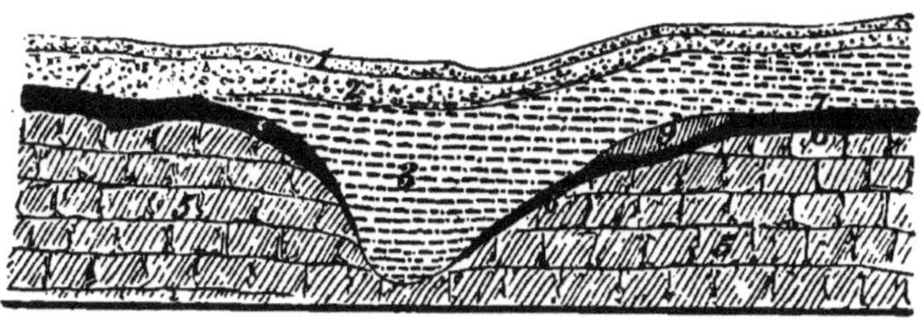

FIG. 67.—CROSS SECTION ALONG LINE A B OF THE SAME.

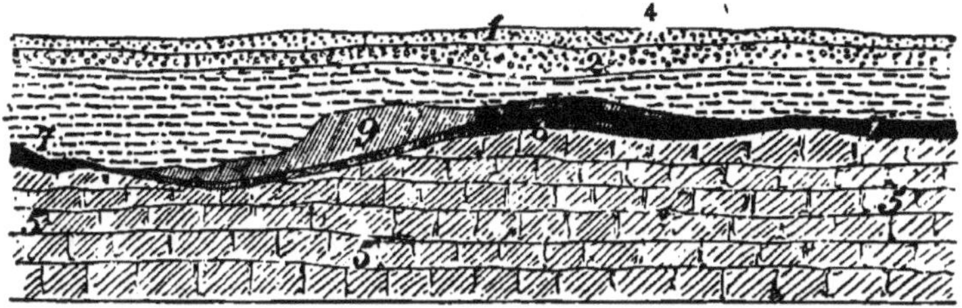

FIG. 68.—LONG SECTION ALONG LINE C D OF THE SAME, SHOWING MANGANESE.
1, Soil. 2, Quartz Drift. 3, Clay. 4, Shale. 5, Limestone. 6, Sand. 7, Manganese. 9, Hæmatite.

In fig. 69 a continuous layer of manganese rests upon an equally continuous layer of mulm or sand, both following the worn edges of the limestone. In fig. 71 we have an example of the thinning out of both mulm and manganese, with the occurrence of brown ironstone, and of the same containing nests of manganese occurring over the manganese.

The thickness of the manganese varies from 6 inches to 15 feet, the average thickness taken from a great number of mines being 1 foot 6 inches.

The average quality of the manganese obtained from these different workings is about 63 per cent. of peroxide of manganese. The quantity obtained from one square lachter = 5 feet × 5 feet × 5 feet, is about 5 tons. Some years since there was computed to be in the district 2,800,000 quadralachters of the mineral; the mines numbered 259, and the production had reached 25,000 tons a year.

Farther north, where the rocks beneath the limestone come to the surface, there are numerous small accumulations of manganese in them, and the ore is widely disseminated through them. I have already noticed that manganese occurs in the limestone itself. In all probability, therefore, these masses of the mineral are due to the same wearing down and washing out processes already described, by means of which the manganese has been extracted from its parent rocks, and accumulated in layers and nests, as we now find. The stiff clay belongs to an age preceding the long and somewhat indefinite period known as glacial.

The mining is very simple. As in the mining for apatite in the same region, small shafts are sunk through to the clay. These are, where the clay is wet or loose, cased with wickerwork. Speaking from experience, I can say that the underground workings are tortuous and intricate, like a rabbitwarren. Little or no timber is used, the clay ordinarily being strong enough to stand without assistance for the time required. Mining operations are confined chiefly to the summer months, partly because of the drier state of the clay then, and also on account of the washing of the ore, which is

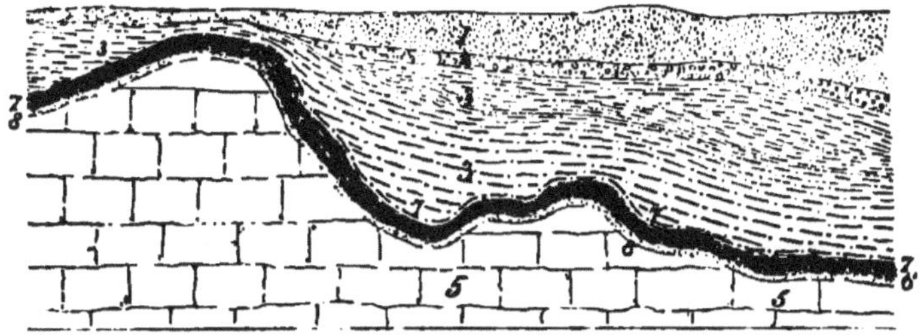

FIG. 69.—SECTION OF THE RUBIN MINE, NEAR NIEDERTIEFENBACH.

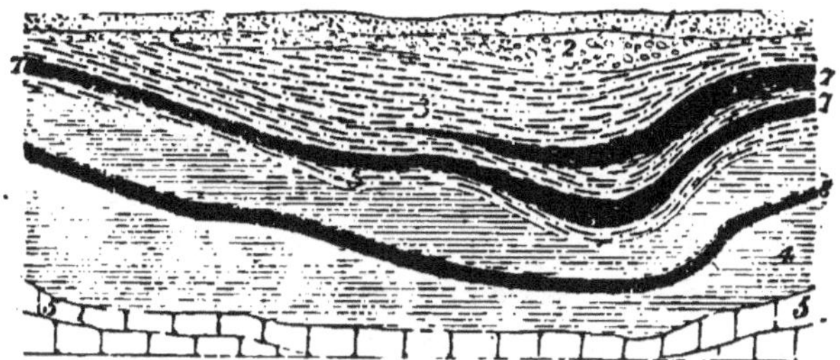

FIG. 70.—SECTION OF THE FAHRWEG MINE, NEAR NIEDERTIEFENBACH.

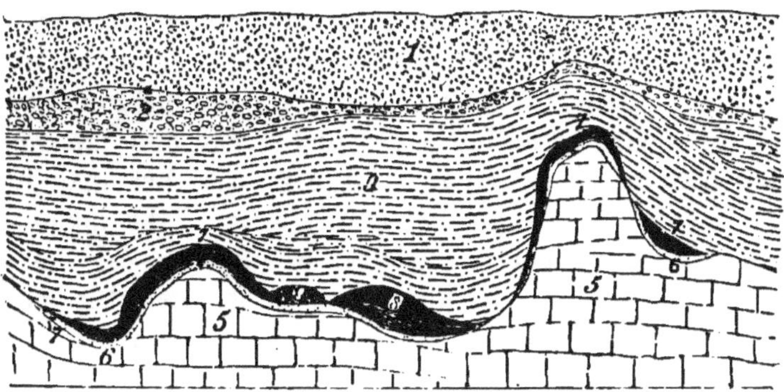

FIG. 71.—SECTION OF THE HÖCHST MINE, NEAR NIEDERTIEFENBACH.

1, Soil. 2, Quartz Drift. 3, Clay. 4, Shale. 5, Limestone. 6, Sand. 7, Manganese. 8, Brown Ironstone with nests of Manganese.

done on tables and on troughs in the open air by youths and young women by the side of the nearest stream—where practicable, by the side of the river Lahn itself.

Cheaply as the mining is conducted, it is difficult, after the freight down the Lahn to the Rhine, the transhipment to Rhine barges, the transhipment to seagoing vessels, and the freights down the Rhine and by sea to England are paid, to make mining manganese in this district for exportation to foreign parts to pay.

ITALY.—A group of manganese mines has been worked for more than a hundred years in the Val d'Aosta and the Val Toumanche, Italy. They seemed to have been worked for other minerals at a much earlier period, it is said by the Romans, who left a large quantity of ore about the mines. The names of this group of mines are, St. Marcel, now worked on a lode; Tourgnon, also worked on a lode; Val Toumanche, worked on an immense irregular deposit; and Bardonecha, worked on a series of nearly horizontal beds, which are about one metre thick. The ores are the ordinary oxides, varied with proportions of sulphur. They frequently occur in a crystalline form, and all of them contain a small proportion of silver.

It is probable that the lodes referred to are simply branchings out from the great irregular masses and beds in which the workings were first prosecuted. Thus Saussure describes the mine of St. Marcel, nearly a century ago, as 'situated on a mountain of gneiss, the beds of which are in a horizontal position. The mine is entirely open to the day, and the ore is supposed to be deposited in a large mass rather than in a bed or a vein. It lies parallel to the strata of the mountain, and it is about 15 feet in thickness where it appears on the surface; but it gradually diminishes as it enters the mountain to the thickness of 5 or 6 feet. It has been penetrated about 50 feet, and it does not appear to exceed 200 to 300 feet in length. Its inclination from the west is at an angle of 15 to 20 degrees. The mine affords fine specimens of the red ore or carbonate of manganese, which are of a beautiful red colour, and crystallized in the form of rhomboidal prisms.'

In a recent statement the costs of working the St. Marcel mine are thus given. Mining, per ton, 15 francs, royalty 5, carriage to highway 30, along highway to railway station 15 = 65 francs. The value of the mineral is put at 4*l.*, so that deducting 65 francs = 2*l.* 12*s.*, a profit of 1*l.* 8*s.* per ton is left. In this statement there is nothing put down for exploration and dead work, nor for management, so that it is probable that in actual experience the margin of profit would be reduced.

SPAIN.—In the mining district of Huelva, Spain, represented on the annexed map, fig. 72, a district famous for its rich pyrites and hæmatite deposits, a very large quantity of manganese has been raised, there being some years since some 150 miners at work. These manganese mines stretch over the south-west part of the province of Huelva, with a portion of the contiguous part of Portugal. The greater part of this region lies about 800 feet above the sea. It is an unfruitful region, the population being chiefly supported by the mines.

The manganese lies in clay slate belonging to the Silurian group of strata (Lower Silurian probably). The rock is of a greyish colour and shining appearance. The general strike of the strata is from north-east to south-west, with a dip of from 40° to 50° to the south-east. Between the villages El Alosno and Castillejos these schists are about a mile long and 2,000 feet in thickness, very compact, and with protrusions of felspathic and greenstone porphyry rocks. The manganese occurs in the clay slate in the vicinity of these harder rocks, and often at the point of contact between them and the clay slate. It is associated with quartzites and small ironstones, of both of which there are layers and nests interstratified between the slates. The manganese occurs in the contact of the slates with these layers. Usually the manganese is not co-extensive with the ironstone layers, but it is of considerable thickness, and it opens out into great nests; usually the thickness is less as the extent is greater. Thus, at the mine of St. Cataline, the deposit extended 1,300 feet with a thickness of 2 feet, while in the

FIG. 72.—MAP OF THE HUELVA MANGANESE MINING DISTRICT, SPAIN AND PORTUGAL.

mine Louisa the deposit was 150 feet in length with a thickness of 25 feet.

In the layers the manganese and iron do not preserve a similar or equal thickness throughout their extent. One mineral presses out the other, so that one or the other often occupies the whole space between the under and overlying rocks. The two minerals, though so closely associated, are not mixed, but are kept distinct from each other, so that usually the iron does not contain manganese nor the manganese iron; a natural arrangement which we have seen prevails elsewhere, and one that much facilitates the working of both minerals.

The principal ores of manganese worked in this district are pyrolusite and psilomelane, wad and manganite seldom occurring. The best analysis shows 94 per cent. of oxide of manganese, the average quality ranging from 70 to 75 per cent. The thickness of the nests of manganese, with the underlying iron, ranges from 10 to 40 feet. The concessions are usually 300 metres long and 200 metres broad, with vertical boundary-lines. The proportion of productive to unproductive ground in these concessions is very variable. The workings were some time ago all open to day, but as the ore is won the layers are followed downwards. The cost of getting and separating 100 lbs. of manganese is stated at from 2 to 3 reals, the minerals hand-picked, and the adhering clay, slate, or other substances separated from it. Manganese ores also occur near the Cape de Gata, the south-east point of Spain, and the southernmost point of the Sierra Algamilla. The mining district is about 35 miles long by 16 miles broad. It has cup-shaped hills of trachyte and porphyry, which contain ores of manganese, lead, and copper. On the west side of these hills there are calcareous shales of Tertiary age, which are upheaved and broken by the trachytic and porphyritic rocks. In these latter rocks there are manganese ores of a quality of 70 to 90 per cent. of peroxides, but the deposits are so small in extent and thickness that it is scarcely possible to make the working of them pay. In the abrasion of these rocks, however, during long ages, the manganese has been separated from them, and has been gathered into nests and layers in the Tertiary strata, which are frequently a calcareous conglomerate. Some of these deposits have an extent of 250 feet in length, 150 feet in breadth, and a thickness of 5 feet. They contain peroxide of manganese of from 75 to 80 per cent. The strike of the beds here is from north-west to south-east and the dip north-east 25 degrees to 30 degrees.

FRANCE.—This country has been rich in manganese. The old mine of Romanèche, in the Department of Seine and Loire, has been worked for this mineral over a hundred years. It is situated on the eastern slope of a chain of mountains running north-north-east and south-south-west, composed of

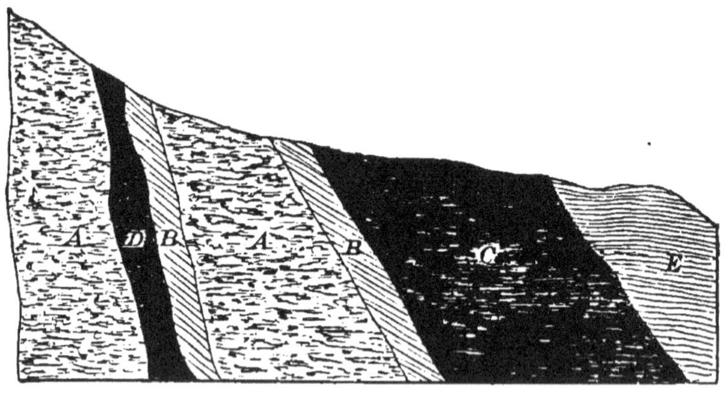

FIG. 73.—SECTION OF THE DEPOSIT OF MANGANESE OF THE OLD MINE OF ROMANECHE, FRANCE.

A A, Granite. B, Limestone. C, Manganese. D, Vein of Manganese. E, Greenish Clay.

granite, limestone, and hard sandstones, as seen by the annexed sketch, fig. 73, which shows how the deposits were understood about a hundred years ago.

There is first of all a thin vein or irregular bed of manganese in the granitic rock, and then, a little higher up, a much larger deposit, the main one, which is surmounted by what is near the surface a greenish clay, but which lower down becomes indurated. This mass was about 400 yards in length, and about 20 yards at its greatest breadth. It extends from north-east to south-west, dying out at the extremities. It forms thus an irregular mass, accumulated in the bedding of the strata. It rises in places above the soil, so that probably it was only a portion of a much larger mass, the upper part of which had been denuded. The greater part of the mass was free from an admixture of other minerals, but in places there occurred fluor spars of a deep violet colour, and the cavities and fissures contained a reddish grey plastic clay. Through the kindness of M. Vital, of Rodez, Aveyron, Government Inspector of Mines for that district, I am able to give a plan of the workings on the two lodes or deposits, fig. 74, and also a detailed description by him of each lode as occurring about the year 1857. The lodes are still worked.

'The lodes explored in the neighbourhood of Romanèche are three in number—the great vein, the little vein, and the mass subordinate to the great vein.

'The *Great Vein* has an easterly direction, north 10° east, and dips to the east at an angle of about 70°; it is situated upon the border of a mass of greatly decomposed granite. To

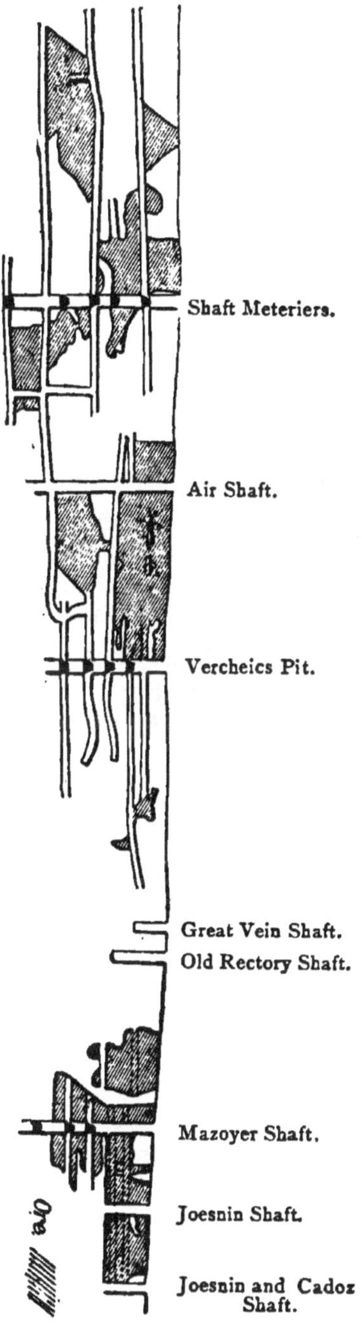

FIG. 74.—LONGITUDINAL SECTION OF THE LITTLE VEIN AND OF PART OF THE GREAT VEIN OF THE ROMANÈCHE MINE, FRANCE. Scale, 1-4000.

the south it distinctly penetrates the granite and ramifies. To the north it is lost in a great network which penetrates the oolitic limestones in contact with the granite. Between these two extremities its length is 320 metres = 349 yards.

'The *Little Vein* branches out from the great vein at 120 metres = 131 yards from its northern end. Its direction is north 35° east, and its dip is from 70° to 72° to the east. It is imbedded in the granite and has been followed about 530 metres = 577 yards towards the south.

'The veins consist of manganese oxide, barytiferous hydrate (prilomelane), quartz, fluor spar, sulphate of barytes, and oxide of iron. These minerals are mixed together in all proportions, and the mixture forms the mineral matter of the lodes.

'The thickness of the Great Vein varies in the regular parts from 1·50 m. to 2 m. = 4 ft. 10 in. to 6 ft. 6 in., and that of the Little Vein from 3 ft. 3 in. to 4 ft. 10 in.

'The portions near the junctions of the lodes were the most mineralised, and near here the Little Vein was about 6 metres thick = 6½ yards, close to the surface.

'The exploratory works have been sunk about 80 metres = 87 yards in the Little Vein, and about 36 metres = 39 yards in the Great Vein. The richness of the veins seems to diminish in depth, and the exploratory works are concentrated in the neighbourhood of the levels. The raw material yields about 40 per cent. of commercial mineral.

'The subordinate mass to the Great Vein consists of a layer nearly horizontal of argillaceous materials mixed with pebbles of a varied nature and of mineral blocks. This stratum runs along the roof of the Great Vein near the surface. It has been followed about 100 metres = 109 yards, about 60 metres = 62 yards wide, and 30 metres = 33 yards thick, and the raw material makes 75 per cent. of commercial mineral.

'The veins were discovered in 1750 and were immediately worked. The mass was discovered in 1847, and since then the principal works have been in it.

'The works supply three qualities of minerals—the rich 65° and above, the medium 54° to 58°, and the poor 45°.

'In 1881 the mines employed 182 men, with an output of 10,870 tons French, equivalent to a value of 365,632 francs.'

In the Cevennes manganese is found under similar mineralogical conditions, and the ore is characterised by being light and friable, and as dividing in irregular prisms.

Manganese is also met with in what were the Departments of the Vosges and the Moselle, and a compact variety of grey ore has been produced in great quantities in the department of Dordogne, which has been known in commerce as Périgueux stone.

In the Emma lode of the La Vidale mine of the Mines d'Asprières, Aveyron, I have just noticed a nice illustration of the dependence upon or alliance with greenstone by manganese ore. As the lode passes through ordinary Silurian slate rocks it contains, in addition to copper and lead, a great thickness of blende; but in its course of about ten yards through a band or dyke of greenstone the blende gives place to black oxide of manganese.

AMERICA.—Manganese ores are extensively raised in the United States. *Pyrolusite* and *Psilomelane* are found at Bennington, Brandon, Chittenden, Irasburg, and Monkton in Vermont, Conway in Maine, Plainfield in Massachusetts, and Salisbury and Kent in Connecticut. *Wad* or *Bog* manganese occurs at Blue Hill Bay, Dover, and elsewhere in Maine, and at Nelson, Gilmanton, and Grafton, New Hampshire; also at the mine La Motte in Missouri. *Manganese spar* is largely worked at Stony Mountain, near Winchester, New Hampshire.

The sections, figs. 75 and 76, taken from Mr. Raphael Pumpelly's report on the iron ores of Iron Mountain and Pilot Knob, Missouri, illustrate the mode of the occurrence of the manganese deposits of that interesting mining region.

These deposits[1] lie in strata that are considerably higher up in the series than those in which the iron ore deposits proper occur, and they are probably of Lower Silurian age.

[1] *Geology of Pilot Knob and its Vicinity.* By Raphael Pumpelly, United States Geological Survey.

Underlying the deposit of fig. 75 there is a bedded rock of fine grain which has in places the appearance of an indurated

FIG. 75.—MANGANESE DEPOSIT ON CUTHBERTSON HILL.
M, Manganese Deposits. D P, Decomposed Porphyry. G P, Granitic Porphyry.

sandstone, in others that of an altered porphyry. It contains numerous broad and flat cavities filled with an ochreous clay.

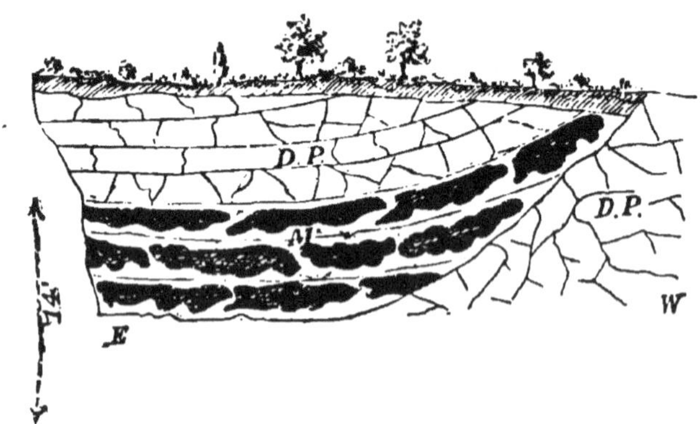

FIG. 76.—SECTION OF MANGANESE DEPOSIT ON BURFORD MOUNTAIN.
M, Manganese Deposits. D P, Decomposed Porphyry.

The manganese in the deposit lies in exceedingly ragged tabular masses. The analysis of this ore is as follows :—

Insoluble silicious matter	0·44
Peroxide of iron	3·30
Manganese as protoxide	68·02
Metallic manganese	52·47

In section fig. 76, at Burford Mountain, which lies west of the above, we have a bedded deposit lying between decomposed porphyry above and below, of a pink colour, with the difference that the ore is a manganiferous iron ore of a very superior quality, and, it is said, is a remarkably fine ore for the manufacture of *Spiegeleisen*. An analysis shows it to contain—

Insoluble	8·54
Peroxide of iron	68·30
Manganese as protoxide	15·84
Sulphur	0·017
Phosphorous acid	0·012

Equal to—

Metallic iron	47·81
Metallic manganese	12·32
Sulphur	0·017
Phosphorus	0·044

These deposits are interesting as showing the similarity in the manner of the deposition of the manganese in the ancient felspathic rocks with that in which the mineral is deposited in the old Tertiary clays of Nassau, Germany, and the clays above the Carboniferous limestone, as shown in fig. 63.

Among the Southern States the manganese mines of Cave Spring, Louisiana, were lately in successful working, producing about 16 tons of ore a day.

Interesting deposits of manganese occur at the 'Lucky Cuss' mine, Tombstone, Arizona. The country rock is a hard white limestone, in which the manganese ore bodies occur in 'pipes' or 'chutes,' which are of great depth, but do not extend far horizontally. One of these is nearly round in shape, 10 to 12 feet diameter. Another is a flat, oval-shaped mass, 40 feet long by 3 broad. Another is of the same breadth but longer. Some of these ore bodies have been followed to a depth of 100 feet without any signs of diminution in size. The ore is a rich oxide of manganese, containing about 10 per cent. of silica and 25 ounces of silver to the ton of ore. Small bunches of galena occur here and there throughout the masses, and these contain up to 200 ounces of silver to the ton of ore. Usually the more silica the manganese contains the richer it is in silver, although an occasional mixture of manganese and lime spar shows a higher percentage. About 22,000 tons of ore had, up to May, 1882, been obtained from the mine at a cost of $10 a ton, the cost getting lower as preparatory work becomes finished.

Small shipments have also been made from Canada. The Queen manganese mine of New Brunswick shipping in May, 1882, a cargo of 105 tons.

From the foregoing particulars it will be seen that originally manganese was largely disseminated throughout the older rocks, into the cracks and veins of which it was subsequently gathered. That, as occurring in veins and lodes, it is not usually in sufficient quantity to pay for working. At a more recent period it seems to have been deposited contemporaneously with the strata, particularly in strata of a sandy nature and in the vicinity of porphyritic and felspathic rocks. The more recent deposits in Tertiary clays are derived from the denudations of these older strata. It will also be seen that from the irregular nature of the deposits the mining must be of the cheapest character and the machinery of the most temporary nature compatible with safety.

CHAPTER XIX.

CLASSIFIED LIST OF MINERAL SUBSTANCES.

Purposes of the Chapter—Abbreviations—List of Simple Elements, divided into Metallic and Non-metallic—Further divided into Seven Classes—Oxygen—Enumeration of Classes with Included Substances—Table of Strata—Conclusion.

THE following list of minerals occurring in nature is intended chiefly to show the position occupied by those minerals which are described in this volume, and in that of *Metalliferous Minerals and Mining*. As far as possible the chemical composition of each mineral is given or indicated. It need hardly be said that differences of opinion exist as to the precise order or succession in which mineral substances should be placed, depending upon an author's preference for one or another mode of grouping, according to chemical composition, or affinities, or upon crystallisation, or upon the behaviour of the various minerals under treatment. In the following list I have followed as nearly as I could the arrangement of Dana. The abbreviations used are:—H. hardness; E. M., *Earthy Minerals and Mining;* M. M., *Metalliferous Minerals and Mining.* The whole of the materials of which, as far as we know, the crust of the earth is formed, is reducible to the sixty-four elementary substances enumerated in the following list:—

LIST OF ELEMENTARY SUBSTANCES WITH THE ABBREVIATIONS BY WHICH THEY ARE USUALLY KNOWN.

Aluminium	Al.	Barium	Ba.	Boron	B.
Antimony	Sb.	Beryllium	Be.	Bromine	Br.
Arsenic	As.	BISMUTH	Bi.	Cadmium	Cd.

LIST OF ELEMENTARY SUBSTANCES—*continued.*

Cæsium	Cs.	LEAD	Pb.	Selenium	Se.
Calcium	Ca.	Lithium	L.	SILVER	Ag.
Carbon	C.	Magnesium	Mg.	Silicon	Si.
Cerium	Ce.	Manganese	Mn.	Sodium	Na.
Chlorine	Cl.	MERCURY	Hg.	Strontium	Sr.
Chromium	Cr.	Molybdenum	Mo.	Sulphur	S.
Cobalt	Co.	NICKEL	Ni.	Tantalum	Ta.
COPPER	Cu.	Niobium	Nb.	TELLURIUM	Te.
Didymium	D.	Nitrogen	N.	Thallium	Tl.
Erbium	E.	Osmium	Os.	Thorium	Th.
Fluorine	F.	Oxygen	O.	TIN	Sn.
Gallium	G.	PALLADIUM	Pd.	Titanium	Ti.
GOLD	Au.	Phosphorus	P.	Tungsten	W.
Hydrogen	H.	PLATINUM	Pl.	Uranium	U.
Indium	In.	Potassium	K.	Vanadium	V.
Iodine	I.	Rhodium	Rh.	Yttrium	Y.
IRIDIUM	Ir.	Rubidium	Rb.	ZINC	Zn.
IRON	Fe.	Ruthenium	Ru.	Zirconium	Zr.
Lantanum	La.				

These elements are broadly divided into metallic and non-metallic, the latter being fifteen in number, and the whole of them may be conveniently divided into the seven classes described in this chapter. It will be noticed that many of these elementary substances are very rare, and, so far, of more interest scientifically than commercially, the most abundant elements being silicon, oxygen, lime, magnesia, sulphur, with others described in this volume. Of oxygen, as it does not find a place among the gases occurring native, Class I., it may be well to say a few words here, unless indeed, from the correspondence it bears to sulphur in the way in which both these minerals combine with certain others, we place it in the same class as sulphur. It has a general relation to the whole of the other elements, all of them, excepting fluorine, combining with it to form oxides. Oxygen was discovered almost simultaneously, but independently, by Priestley and Scheele, Priestley in 1774 and Scheele in 1775. In 1778 Lavoisier described the position occupied by oxygen in the atmosphere, and showed the changes that took place when bodies burn in the air. He gave it the name of oxygen, from *oxus*, acid, and *gennao*, I generate, with reference to its property of forming acids in uniting with other elementary bodies. Combined with these, it is the most extensively diffused and abundant substance in nature, forming part of the atmosphere, of water, and of nearly all the substances of which the globe is composed. Its discovery is also taken as the date of the origin of true chemical science and theory. It may be obtained from the atmosphere as well as from other substances, but the material most used in its manufacture is the black oxide of manganese (page 287). Oxygen gas is without colour, smell, or taste. It is heavier than air; the latter being 1,000, oxygen is 1,102·6. It is essential to life and combustion, and combustion takes place with greater brilliancy and swiftness where it is present in excess.

The degree of hardness assigned to the various minerals by the figures used in the description given of them will be understood by a reference to the following scale:—

1. Talc.
2. Rock Salt.
3. Calcareous Spar.
4. Fluor Spar.
5. Apatite.
6. Adularia Felspar.
7. Rock Crystal.
8. Prismatic Topaz.
9. Corundum.
10. Diamond.

It will be noticed in the perusal of the following list and of the detailed description of substances given in this and the volume on metalliferous minerals, how recently, comparatively speaking, the true nature of many of them has been discovered; that, notwithstanding the rapid progress which the sciences of chemistry and mineralogy have made during the last hundred and fifty years, they may still be taken, if not in their infancy, at least in a state of growth and progression.

CLASS I.—GASES.

The gases are divided into—

1. Those consisting of or containing nitrogen, atmospheric air.

2. Those consisting of or containing hydrogen, carburetted hydrogen, phosphuretted hydrogen.

3. Those consisting of or containing carbon or sulphur, carbonic acid, sulphurous acid.

Atmospheric Air.—Composition: oxygen 21 per cent. by weight, and nitrogen 79 per cent., with a small quantity of carbonic acid. About 815 times lighter than water. Essential to life. The oxygen consumed by the breathing of animals and consumption of fuel is given back by vegetation. Pressure, about 15 lbs. to the square inch. Encircles the earth to a height of about 45 miles above the sea.

Nitrogen.—A colourless gas, without taste or smell, is lighter than air, as air being 1, its specific gravity is 0·972. Is destructive to life, not by poisoning, but by suffocation. It does not combine readily with other substances, but it can be made to combine with oxygen and with hydrogen. When combined with the latter it forms ammonia. It may be prepared by passing air over red-hot copper, or by passing a current of chlorine through strong solutions of ammonia. Although of itself unable to support life or combustion, it forms a large proportion of the air we breathe. It is freely given off from various warm mineral springs.

Carburetted Hydrogen.—Hydrogen does not exist uncombined in nature. *Carburetted Hydrogen* consists of carbon 75 and hydrogen 25. It is nearly identical with the gas in ordinary use for lighting purposes, and it issues freely, and, as is too well known, with terrific results, from coal seams.

Phosphuretted Hydrogen.—Composition: phosphorus 91·29, and hydrogen 8·71. Supposed to be the same as the luminous matter seen hovering over bogs and marshes.

Sulphuretted Hydrogen.—Composition: sulphur 94·2, hydrogen 5·8; of a putrid taste and smell. Common about sulphur springs and volcanoes.

Muriatic Acid, Hydrochloric Acid.—Composition: chlorine 97·26, hydrogen 2·74. Largely made for manufacturing purposes. Is pungent in smell, and acrid to the skin.

CLASS II.—WATER.

Water was first shown by Cavendish, in 1781, to be the product of the combustion of hydrogen and oxygen. Humboldt and Gay-Lussac afterwards demonstrated that the two gases unite strictly in the proportion of two volumes of hydrogen to one of oxygen, and that the water produced by their combinations when in a state of vapour occupies two volumes. A series of experiments subsequently established the proportions by weight as hydrogen one part, and oxygen eight parts. Specific gravity 1. At 32° it freezes, and as ice its specific gravity is lighter, being 0·916. In this form it assumes a blue or greenish colour. It crystallises in a rhomboidal form. As snow it crystallises in a variety of combinations. Its density is greatest at 39·1, and at temperatures below this it expands. It boils at 212°, but its exact boiling-point is varied according to its degree of purity and the nature of the substances held in solution. It contains various proportions of atmospheric air, in which the proportion of oxygen is higher than in the air itself. In sea-water, there are solid substances amounting from 32 to 37 parts to 1,000. Of these, usually more than one-half is common salt, and four-fifths of the remainder magnesian salts, with sulphate and carbonate of lime, and traces of bromides, iodides, phosphates, and fluorides. These are most abundant in the Atlantic, and least in the Baltic Sea. An analysis of the water of the British Channel gives to 1,000 parts: water 964·7, chloride of sodium 27·1, chloride of potassium 0·8, chloride of magnesium 3·7, sulphate of magnesia 2·30, sulphate of lime 1·4, carbonate of lime 0·03, with traces of the other minerals mentioned above. For other analyses of sea-water, see pp. 83, 90, 91.

CLASS III.—EARTHY MINERALS.

1. Silica and its varieties. See pp. 3—15.
2. Lime and its varieties. See pp. 31—35.
3. Magnesia and its varieties. See pp. 24—28.
4. Alumina and its combinations. See pp. 16—24.

5. Glucina and its combinations. See p. 35.
6. Zirconia and its combinations. See p. 36.
7. Thoria. See p. 36.

Beryllium, probably identical with glucinium.

CLASS IV.—CARBON AND COMPOUNDS OF CARBON.

For the description of carbon and its combinations, diamond, graphite, jet, amber, bitumen, petroleum, &c., see p. 183.

CLASS V.—SULPHUR.

See p. 232.

Selenium.—Selenium is allied to sulphur, and we have seen how it is present, together with that mineral, in several bodies. It was discovered by Berzelius in the year 1817, in the sulphur of the copper mine of Falun, which was employed in a sulphuric acid manufactory in Sweden. It is one of the rarest elements, but it occurs in minute quantities in several of the ores of copper, lead, silver, bismuth, tellurium, and gold; found in Norway and Sweden, and also in the mines of the Hartz in Germany, as well as in the Sipan Islands. The process by which it is separated from its combinations with other substances is a very complicated one. It is, or was, extracted from a seleniferous ore of silver in Norway, and sold in little cylinders about three inches long, of the thickness of a goose-quill.

CLASS VI.—HALOID MINERALS.

(Compounds of the earths and alkalies), with notices of some of their metallic bases.

Ammonia.—Ammonia takes its name from Ammonia in Libya, where a salt was extracted, named after the region sal ammoniac, and from which ammonia has usually been obtained. In a state of purity ammonia is a gas, of which the liquor is a solution in water. Ammonia is produced by the destructive distillation of organic matters containing nitrogen. For, as we have already seen, it is a combination of nitrogen and hydrogen.

Sal Ammoniac.—Composition: ammonia 33·7, chlorine 66·3. Formerly obtained largely in Africa from soot of the fires made from the dung of camels; obtained also from bones and hoofs and horns, and latterly from the ammoniacal liquor obtained in the making of coal gas. It has varieties—

Mascagnine, Sulphate of Ammonia.—Composition: sulphuric acid 53·3, ammonia 22·8, water 23·9; yellowish grey or lemon yellow colour.

Phosphate of Ammonia, found in guano.

Struvite, a phosphate of ammonia and magnesia.

POTASH OR POTASSA. COMPOUNDS OF POTASSIUM.—*Potassium*, one of the simple elements, produced from potash, in 1807, by Sir Humphrey Davy. It has, when first produced, a white colour with a shade of blue. It is solid at an ordinary temperature, but yields like wax under pressure. It oxidises immediately on exposure to the air, and soon loses its colour, and is covered with a dull film of oxide. It is brittle at 32°, and at this temperature has crystallised in cubes; at 70° it is semi-fluid, and becomes liquid at 150°. It can be distilled at a low red-heat, when it forms a green-coloured vapour. At 60° its specific gravity is ·0865. It appears to have, of all bodies, the greatest affinity for oxygen. Exposed in thin slices to the atmosphere it passes into a white matter, which is the protoxide of potassium, or POTASH.

Nitre, Nitrate of Potash, Saltpetre.—Composition: potash 46·56, nitric acid 53·44. Occurs in India, and appears as an efflorescence from the soil in Egypt and Spain, where considerable quantities are collected. It is artificially produced in various European countries in *nitriaries*, or nitre-beds, from the decomposition of the nitrates of lime and magnesia, which are common in the neighbourhood of the beds, and also from refuse animal and vegetable matter.

Chloride of Potassium, Sylvine.—Occurs with salt at Salzburg.

RUBIDIUM, CÆSIUM.—These two metals resemble each other and potassium so closely that prior to the year 1860—1, they were mistaken for that metal. About the time stated they were discovered by Bunsen and Kirchhoff, by means of spectrum analysis. They were first detected in the water of Durkheim, but subsequently they were found in many springs and micaceous and in other of the older rocks. Rubidium is a white metal, which rapidly oxidises and gives off a green vapour. Both together they are separated from potassium, cæsium being subsequently separated from rubidium.

SODA.—Soda is a combination of sodium (see p. 61) with hydrogen and oxygen. It comprises the following salts, which are all more or less soluble, in which they differ from those of potash. H. under 3; specific gravity under 2·9.

Sulphate of Soda, Glauber Salt.—Composition: soda 19·3, sulphuric acid 24·8, water 55·9. White to yellowish white; taste, saline and bitter. Differs from Epsom salts in its coarser crystals and yellow colour under the blow-pipe. Occurs in a cave on Hawaii, one of the Sandwich Islands, and is prepared from sea-water. First discovered by a German chemist named Glauber.

Nitrate of Soda. See p. 100.

Natron, carbonate of soda.—Composition: soda 2·18, carbonic acid 15·4, and water 6·28; but usually mixed with chloride of sodium, and other salts. Occurs in Egypt in the soda lakes, and in the valley of Bahr-bela-ma, thirty miles west of the Delta. A variety named *Trona* occurs between Tripoli and Fezzan, in Africa, where it forms a thin layer under the soil, which yields several hundred tons yearly. Is also found in saline lakes associated with common salt.

Chloride of Sodium, common salt. See p. 61.

Borate of Soda, Borax. See p. 102.

LITHIA.—Is a rare alkaline oxide, whose metallic base, LITHIUM, was discovered by Arfwedson in the year 1818. The metal was obtained by Sir Humphrey Davy by the voltaic decomposition of lithia. It is white, resembling sodium, and is very oxidisable.

BARYTA.—For *Barium*, *Baryta*, and their combinations, see p. 103.

STRONTIA.—*Strontium* is one of the simple elements. It

is a white metal, denser than oil of vitriol. It resembles barium; it has not a high lustre, is fusible with difficulty, and is not volatile. It was first obtained by Sir Humphrey Davy in 1808. By exposure to the air, or by contact with water, it is changed to *Strontia*, which consists of—

Strontium	84·54	. . . or 1 atom 44
Oxygen	15·46	. . . or 1 atom 8
	100·0	

Strontia derives its name from Strontian, a mining village of Argyllshire. It is not abundant in nature.

Sulphate of Strontia, Celestine.—Composition: strontian 56·4, sulphuric acid 43·6. H. 3· to 3·5. Specific gravity 4. Brittle, columnar, crystallises in rhombic prisms, clear white, with a tinge of blue. Used in the manufacture of fireworks for producing a red colour.

Carbonate of Strontia, Strontianite.—Composition: strontia 70·2, carbonic acid 29·8. H. 3·5 to 4. Specific gravity 3·6 to 3·72. Greenish white, grey, and yellowish brown. First distinguished from carbonate of barytes in 1790 by Dr. Crawford. Occurs at Strontian in starlike and fibrous groups, associated with galena. Also used in the manufacture of fireworks.

LIME.—See p. 31 and p. 109 *et seq.*, apatite or phosphate of lime.

MAGNESIA SALTS.—*Sulphate of Magnesia, Epsom Salts.*—Composition: magnesia 16·3, sulphuric acid 32·5, water 50·2. Occurs in fibrous crusts or botryoidal masses of a white colour, also in fine small rhombic crystals. Found at Epsom in Surrey, at Seidlitz, and various places in Europe; in the Cordilleras of Chili, also in South Africa.

Magnesite, Carbonate of Magnesia.—Composition: magnesia 47·6, carbonic acid 47·6. Occurs in fibrous plates and in minute acicular crystals of a white, yellow, or grey colour. Found in magnesian limestones, and used sometimes for the manufacture of sulphate of magnesia.

Brucite.—Composition: magnesia 69·0, water 31·0. After

exposure it often contains carbonic acid. Colourless to greyish white. Translucent, pearly, soluble in acids.

Nemelite is a fine fibrous variety with a silky lustre. Composition: magnesia 62·0, protoxide of iron 4·6, water 28·4, carbonic acid 4·1. Resembles asbestos.

Boracite, Borate of Magnesia.—Composition : magnesia 30·0, boracic acid 7·0. White or grey colour, with a vitreous lustre. H. =7. Specific gravity 2·97. Occurs with gypsum and common salt. Other minor varieties are—

Nitrate of Magnesia.—Occurs in limestone caverns, associated with nitrate of lime.

Polyhalite, brick-red in colour, composed of sulphates of lime, potash, and magnesia, with water.

Rhodizite, like boracite, but tinges the blow-pipe flame red. Found in Siberia with red tourmaline.

Wagnerite.—Composition : phosphoric acid 43·32, fluorine 11·35, magnesia 37·64, and 7·69 magnesium, varied with 3· to 4·5 iron peroxide, and 1 to 4 lime.

See also p. 24, *et seq.*

ALUMINA, SALTS OF.—1. *Native Alum.* See p. 107.

2. *Alunite,* alum stone.—Composition : alumina 37·1, sulphuric acid 38·5, potash 11, water 13. Colour white, greyish, or reddish, vitreous lustre. Crystals rhombohedral, transparent to translucent. H. = 4. Specific gravity 2·58—2·75. A variety found in Hungary is hard enough for the manufacture of millstones.

Websterite, another form of sulphate of alumina, also called alumnite.

Wavellite.—Composition : alumina 33·8, phosphoric acid 34·9, water 26·6, fluoride of aluminium 4·6. Occurs in small half-round masses adhering to surface of rocks, of a white or yellowish or brown colour, with a pearly lustre. *Fischerite* is closely allied ; has a dull green colour.

Turquoise.—Composition : alumina 44·5, phosphoric acid 30·9, oxide of copper 3·7, protoxide of iron 1·8, water 19·. Occurs in reniform masses ; colour bluish green, with a waxy lustre. H. = 6. Specific gravity 2·6—3. Found in a mountainous district near Nichabour in Persia; said to occur in veins which traverse the mountains. Receives a fine polish, and is much valued as a gem.

Hydrate of Alumina, Gibbsite.—Composition: alumina 65·6, water 34·4, with traces of phosphoric acid. Is softer than chalcedony, which it resembles. Other varieties are—

Amblygonite.—Composition: phosphoric acid, alumina, and lithia.

Childrenite.—Composition: phosphoric acid, alumina, and water.

Chiolite resembles cryolite.

Cryolite occurs in snow-white masses, melts in the flame of a candle; is a fluoride of aluminium and sodium, is quarried to a considerable extent in Greenland, and is used as an ore of aluminium.

Diaspore, or dihydrate of alumina. Found in granular limestone in the Ural Mountains, where it occurs in irregular lamellar prisms with a fine cleavage.

Fluellite.—Composition: fluorine and aluminium. Occurs in the mines of Cornwall in minute forms.

Lazulite.—Composition: alumina 35·7, magnesia 9·3, silica 2·1, protoxide of iron 2·6, and water 6·1. Occurs in compact masses, and occasionally in oblique crystals of a fine azure blue colour, nearly opaque, with a vitreous lustre. Found in clay-slate at Salzburg, and elsewhere.

Mellite, or honey stone, has a resinous appearance, of a honey yellow, and may be cut with a knife. Composition: alumina 14·32, mellic acid 40·53, and water 45·15. A rare mineral, found in Bohemia, Thuringia, and Moravia.

Associated with the haloid minerals just enumerated are the elements CHLORINE, BROMINE, IODINE, and FLUORINE. I have referred to chlorine at p. 61, and fluorine at p. 106, of this volume.

BROMINE.—Was first observed in the year 1826 by M. Balard, of Montpellier. Its name is significant of its bad and disagreeable smell. It is found in very small proportions in sea-water, in the form of bromite of sodium or magnesium;

also in the water of the Dead Sea, and in saline springs generally. The principal source whence bromine, as an article of commerce, is derived, is the wells of Theodorshall, near Kreuznach, Germany. It is closely allied to chlorine in many of the properties of the latter.

IODINE.—M. Courtois, of Paris, in preparing carbonate of soda from kelp, discovered this substance in the year 1811. Its chemical properties were subsequently made known by Clement, Davy, and Gay-Lussac, and it has formed a valuable contribution to medical or surgical resources. It is found in sea-water, but more abundantly in seaweed and sea plants generally. It has also been obtained from an ore of silver at Albaradon, in Mexico.

CLASS VII.—METALS AND METALLIFEROUS MINERALS.

CERIUM.—This name was given to the metal by Hisinger and Berzelius from Ceres. It is not an abundant metal, neither is it used in the arts. It is found in a number of minerals near the celebrated mines of Falun, in Sweden. It has been obtained as a powdery mass of a dark chocolate brown colour, which gave a grey metallic trace under the burnisher.

YTTRIUM.—Is only known to exist with oxygen in YTTRIA, an earth discovered in 1794 by Professor Gadoline, near Ytterby, in Sweden. YTTRIA is white in colour, but generally tinged by the presence of manganese. It is insoluble in water, and infusible, except at a great heat. When dissolved in muriatic acid it gives out chlorine gas, which fact is taken as shewing one property of a metallic oxide.

LANTANUM.—The oxide of this metal was discovered about fifty years since in the cerite of Bastnäs in Sweden, by Mosander, and said to form two-fifths of the oxide of cerium, which, by the ordinary process, is extracted from that metal. The abstraction of the new metal alters but little the properties of cerium, and it lies, so to speak, concealed in that mineral, and it is this circumstance that led M. Mosander to give it the name *lantanum*.

TANTALUM.—This metal appears to have been first discovered in the year 1801, in a black-coloured mineral belonging to the British Museum, which was supposed to have been brought from Massachusetts, and which had been named *Columbian.* In the following year M. Ekeberg found a new metal, which he named tantalum, in two Swedish minerals which he named *Tantalite* and *Yttrotantalite.* Ekeberg gave the name tantalum to the new metal on account of the insolubility of its oxide in acids, in allusion to the legend or fable of Tantalus. The mineral was found to be identical with that of the British Museum named *Columbian.* Berzelius obtained tantalum in the form of a black powder which could be washed and dried, and which gave an iron-grey metallic lustre under the burnisher. It burned in the air below a red heat, and yielded *Tantalic Acid,* in which state it is present in several minerals, especially iron and manganese.

The above four metals are of little or no use in the arts, and are chiefly interesting in a scientific point of view. Although found in minerals in other countries, notably North America, it will be seen that they occur chiefly in Sweden, where their nature seems to have been first accurately ascertained. They seem to be intimately associated one with the other, and altogether with other minerals, in which combinations the following varieties have been observed.

YTTROCERITE.—Composition: oxide of cerium 18·2, yttria 9·1, fluoric acid 25·1, and lime 47·6. It occurs in a massive form of a violet blue colour, ranging to grey and white. H.= 4—5. Specific gravity 3·4—3·5. It is found in Finbo and Brodbo, near Falun, in Sweden, and in Massachusetts and New York, United States of America. Its varities are—

Fluocerine . .	Composition: peroxide of cerium 84·21, hydrofluoric acid 10·85, water 4·95. Yellow to brown, vitreous or resinous. From Finbo, Sweden.
Lantranite . .	Composition: lantanum oxide 52·9, carbonic acid 21·1, water 26. White or yellowish; granular, earthy, and in small tabular crystals. Bastnaes in Sweden, and Sehigh in Pennsylvania.

Parisite . . .	Composition: protoxide of cerium 60·0, carbonic acid 23·6, fluoride of calcium 11·51, water 2·4, with lantanum and didymium. Brownish yellow to red. Found in the emerald mines of the Musso valley in New Granada.

MONAZITE.—Composition: cerium protoxide 25·0 to 37·0, lanthanium oxide 23·0 to 27·0 (18 thorina), phosphoric acid 28, tin oxide 2, lime 1·5, with small portions of manganese and magnesia. Flesh red to reddish brown, with a vitreous lustre. H.=5. Specific gravity=4·8—5·1. Resembles sphene, an ore of titanum, but is distinguished by its brilliant, easy, transverse cleavage. Found at Miask, in the Urals, and in Connecticut and New York States, in North America.

CRYPTOLITE, which is a combination of phosphorus with the oxide of cerium, occurs in minute six-sided prisms, in connection with the apatite deposits of Norway described in Chapter VII.

ALLANITE.—Composition: variable, but usually containing silica 30·0 to 35·0, alumina and iron peroxide 12·0 to 18·0, protoxides of cerium 11 to 24, lanthanium 2·0 to 8·0, iron 4·0 to 21·0, manganese 0· to 3·5, lime 2· to 12·0, yttria 0·3 to 4·0, and magnesia 0·4 to 5·0. In colour black or brown, with an imperfect metallic lustre and a green or greenish-grey streak. H.=6. Specific gravity 3·2 to 3·7. Found in Greenland, Norway, Sweden, the Urals, and in various localities of the United States of America. Its allied minerals or varities are—

Bodenite . .	An ore of cerium resembling orthite. From Boden, Saxony.
Cerine . . .	Consists of silica and alumina, with the oxides of iron, cerium, lantanum, and also lime.
Cerite . . .	Composition: protoxide of cerium with didymium and lantanum 72·0, silica 22, water 6, with iron protoxide and lime.
Orthite . . .	Similar in composition, but occurring in acicular crystals.
Pyrorthite . .	An impure orthite containing carbon, and will burn, hence its name, *pyr*, fire, and orthite.

These varieties are found in the same localities as allanite.

PYROCHLORE.—Composition: very complex. Analyses of the mineral from Miask in Russia, gave niobic acid mixed with titanic and tungstic acid 62·0 to 67·0, lime 10·0 to 13·0, oxide of cerium and thoria 6·0 to 13·0, fluoride of sodium 7·0. In a sample from Norway, yttria, iron, zirconia, lithia and uranium occur. Found near Brevik and Fredericksvärn, Norway, and near Chesterfield in Massachusetts. Minor varieties of the foregoing combinations of cerium, yttrium, lantanum and tantalum with each other, and with other minerals, are the following:—

Æschynite . .	A combination of titanium with zirconia and cerium, in brown and black crystals. Miask, in the Urals.
Euxenite . .	A columbate of yttria with titanic acid and oxide of uranium; brownish black with reddish streak; occurs in splinters. Arendal in Norway.
Fergusonite .	Similar in composition to euxenite, but crystallising in secondaries to a square prism. Cape Farewell, Greenland.
Gadolinite . .	Composition: varied; usually yttria 36·0 to 51·0, silica 25·0 to 29·0, protoxide of cerium with lantanum 5·0 to 17·0, glucina 0·0 to 12·0, protoxide of iron 10·0 to 15. Black with greenish grey streak; vitreous. Kragerö in Norway, and Finbo and Ytterby, Sweden.
Polycrase . .	Like the next following, polymignite: black, massive, and in thin linear crystals; occurs with orthite. Hittero in Norway.
Polymignite .	Composition: protoxide of cerium 5·00, yttria 11·50, zirconia 14·14, peroxide of iron 12·20, peroxide of manganese 2·70, lime 4·20, titanic acid 46·30. Occurs in long prismatic crystals, broad and striated vertically. Iron black, dark brown streak, semi-metallic appearance. Fredericksvärn in Norway.
Rutherfordite .	Contains about 58 per cent. of titanic acid with 10 per cent. of lime, with portions of cerium and yttrium. Is found in the gold mines of Rutherford County, N.C., United States of America.
Samarskite . .	Composition: niobic acid 56·0, protoxide of iron 15·0 to 16·0, oxide of uranium 14 to 17·0, yttria with lime and magnesia 8·0 to 11·0. Velvet black in colour, with a strong metallic appearance. Ilmen Mountains, Siberia.
Tschefkinite .	Similar to Gadolinite. Velvet black; a silico-titanate of cerium. Ilmen Mountains.

Xenotime . .	Composition: yttria 62·82, phosphoric acid 37·18, with 3·39 phosphate of iron, and traces of fluoric acid. Crystallises in square prisms; yellowish brown. Lindesnaes and Hitterö in Norway, and Ytterby in Sweden; also in Georgia, United States of America.
Yttrotantalite .	Composition: tantalic acid 57·9 to 60·0, tungstic acid 1·0 to 8·0, yttria 20·0 to 38·0, lime 0·5 to 6·0, uranium peroxide 0·5 to 6·0, and peroxide of iron 0·5 to 3·5. Occurs in indistinct four or six-sided prisms, and also in grains and lamellæ; in thin splinters translucent; has three varieties in colour: dark or brown, yellow or yellowish grey, and black. Is found at Ytterby in Sweden.

URANIUM.—This is a rare metal in nature. It is of a steel white colour, and in dry air it does not oxidise on exposure at ordinary temperatures. It is derived from *Pitchblende* and *Uranite.* It is used in glass-making, the uranous oxide imparting a fine black and the uranic oxide a beautiful yellow to the glass. In enamelling they yield a fine orange colour. Compounds of uranium are now also used in photography. We have also seen how its oxides enter into the composition of the minerals just described. As just stated its ores are—

PITCHBLENDE.—Composition: protoperoxide of uranium 84·78 and oxygen 15·22, but varied by proportions of lead, magnesia, iron, arsenic, silica, with occasionally vanadium and selenium. Colour, grey brown to velvet black, with a dull or sub-metallic lustre. H. 5·5. Specific gravity 6·47. Dissolves slowly when powdered in nitric acid. Found in the tin mines of Cornwall, in the lead and silver mines of Erzebirge, also in Connecticut, United States of America. Its varieties are—

Coracite . . .	Resembling pitchblende, found near the junction of trap and syenite on the north shore of Lake Superior.
Eliasite . . .	An ore of similar description, but containing 10½ per cent. of water.
Uranic ochre .	A peroxide of uranium of a light yellow colour, easily powdered; associated with pitchblende in the localities named.

URANITE, Uran Mica.—Composition: peroxide of uranium

62·6, phosphoric acid 15·5, lime 6·2, water 15·7; bright yellow to green, with a pale yellow streak. Occurs in short square prisms. H.= 2, 2·5. Specific gravity 3—3·6. Fine examples of green crystalline forms occur in Cornwall. Found also in the mines of the Erzgebirge and near Autun and Limoges, France. Its varieties are—

Johannite or *Uranvitriol* .	A sulphate of uranium of an emerald green colour.
Samarskite, also called *Urantantalite* and *Yttro Ilmenite*	Composition: oxide of uranium 14·0 to 17·0, yttria with lime and magnesia 8·0 to 11·0, protoxide of iron 15 to 16, niobic acid 56, with portions of lime and magnesia. See also p. 323.

Titanium.—In a metallic state sometimes found as small cubic crystals, of a bright copper colour, adhering to the slag of iron furnaces. Their density is 5·3 and they are harder than quartz. As a metal, it was discovered by McGregor of Cornwall in the year 1791. It was afterwards observed by Klaproth, who gave it the name of titanium. In nature it occurs in combination with oxygen, and so combined its ores are with difficulty fusible before the blow-pipe. The ores are—

Rutile.—Composition, titanium 61·0, oxygen 39, sometimes containing a little iron, when it has been called *Nigrine*. Of a reddish brown to red colour. Occurs in eight, ten, and twelve sided prisms. H.= 6—6·5. Specific gravity 4·15 to 4·25. Is largely associated with the apatite deposits of Norway in gneissic rocks (Chapter VII.), with specular iron at the Grisons. Found also at Yriex in France, the Urals, Brazil, Scotland, and the United States of America, sometimes beautifully associated with quartz, p. 7. Used in colouring porcelain. Its varieties are—

Anatase . . .	Same composition as rutile. Occurs in slender octahedrons of a brown colour, and nearly transparent.
Brookite . . .	Same composition as rutile, but with 1·4 to 1·5 of peroxide of iron. Occurs in thin brown hair-like crystals; said to be found near Snowdon and Tremadoc, North Wales.

SPHENE.—Composition: titanic acid 40·4, silica 31·3, lime 28·3, with 0· to 5· of protoxide of iron in the darker varieties; brown, yellow, green, opaque to semi-transparent, with resinous or adamantine lustre. H.=5—5·4. Specific gravity, 3·2—3·6. Occurs in crystals from a quarter of an inch to two inches long in granitic gneiss, mica, syenite, or granular limestone. Found in several localities in Scotland, Norway, Sweden, Saxony, Russia, France and America. The name is taken from the Greek *sphen*, wedge, in allusion to the shape of the crystals. Its varieties are—

Greenovite . .	Containing much protoxide of manganese, and of a flesh red colour. St. Marcel in Piedmont.
Keilhauite or *Yttro-titanite*	Composition: titanic acid 27·8, silica 28·8, lime 19·5, yttria 9·3, alumina 6·9, peroxide of iron 7·7. Of a blackish brown colour with a reddish tinge, and grey brown streak, and vitreous or resinous lustre. Occurs near Arendal, Norway.
Schorlomite .	Of a similar composition to sphene, but of a black shining appearance. From the Ozark Mountains, Arkansas, where it occurs near *gadolinite*.
Perofskite . .	A titanite of lime.
Pyrrhite. . .	Is a niobiate of zirconia with iron and uranium. Found near Mursinsk in Siberia, and at the Azores.
Titantite . .	Is simply a dark variety of *sphene*.

Warwickite occurs in prismatic crystals of an iron grey to brown colour, and a red tarnish. It is distinguished by containing 20 per cent. of boracic acid.

By referring back to the description of the ores zirconia and yttria, it will be seen that titanium enters into the composition of *Æschynite*, *Ærstedite*, *Polymignite*, and others; also of ilmenite—titanic iron.—M. M., p. 248.

NIOBIUM.—Niobium is a rare metal, whose properties are not yet much understood. It seems allied to or associated with tantalum and titanium.

TIN.—For a description of tin and its ores see M. M., pp. 162—186.

ARSENIC.—See pp. 253—257.

COBALT.—See pp. 258—271.

ANTIMONY.—See pp. 275, 281.

BISMUTH.—See M. M., pp. 280—1.

TUNGSTEN.—Tungsten, Swedish heavy stone, as tungstic acid is present in several minerals. In combination with iron and manganese it forms *Wolfram*, or tungstate of iron and manganese (M. M., p. 249); in combination with lead it forms tungstate of lead (M. M., p. 189), and with lime tungstate of lime. Composition: tungstic acid 7·8, lime 19·06; tungstic ochre, forming a yellow powder on other ores of tungsten. It is present also in the minerals pyrochlore and yttro-columbite.

MOLYBDENUM.—See pp. 272—276.

TELLURIUM.—See M. M., p. 286.

LEAD.—See M. M., pp. 187—237.

THALLIUM.—This metal was discovered as recently as the year 1861 by Mr. Crookes in the matter deposited in the flue of a pyrites burner. In a metallic state it closely resembles lead in its physical properties. When fresh cut the newly-exposed surface shows a bluish white lustre, which quickly tarnishes upon exposure, and it is best preserved under water. It is soft, being easily indented with a finger-nail, and it melts below a red-heat. The presence of the metal is shown by the appearance of a vivid green line on the spectrum. It dissolves readily in nitric and sulphuric acids, giving out hydrogen. The soluble salts of the metal are colourless, and act as strong poisons.

VANADIUM.—Is a metal of the rarest occurrence in nature. It was discovered in iron prepared from the iron ore of Taberg, Sweden, in the year 1830 by Sefström, and it was afterwards obtained in larger quantity from the slag of the ore. In its properties it bears considerable resemblance to chromium. As vanadic acid it occurs in—

Vanadinite, or vanadate of lead (M. M., p. 189), discovered by Mr. Johnson, of Wanlockhead.

Vanadate of Copper, found in the Ural Mountains.

Vanadate of Lime, of a brick-red colour, shining lustre, and a foliated structure.

CHROMIUM.—Was discovered in the mineral now known as *Chromate of Lead* (M. M., p. 189) in the year 1797 by Vauquelin. It has since been obtained from other minerals, more especially chromate of iron (M. M., p. 249), and it is from this ore chiefly that the many beautifully coloured preparations used in arts and manufactures are obtained. It forms two compounds with oxygen—oxide of chromium and chromic acid.

NICKEL.—For nickel and its ores, see M. M., pp. 281—284.

MANGANESE.—See pp. 282—308.

IRON.—See M. M., pp. 246—276.

ZINC.—See M. M., pp. 238—245.

CADMIUM.—This is a very rare mineral, of which only one ore is known, named Greenockite. It is of a honey or orange yellow colour, high lustre, and nearly transparent. It is found at Bishopstown in Renfrewshire. Composition : cadmium 77·6, sulphur 22·4. Cadmium is also associated in small proportions with blende and calamine, ores of zinc.

INDIUM.—A metal recently discovered by means of spectrum analysis, associated with zinc ores. A soft white metal resembling cadmium.

MERCURY.—See M. M., pp. 277—280.

COPPER.—See M. M., pp. 114—161.

SILVER.—See M. M., pp. 81—113.

GOLD.—See M. M., pp. 32—80.

PLATINUM.—See M. M., pp. 284, 285.

PALLADIUM.—See M. M., p. 285.

IRIDIUM.—See M. M., p. 285.

OSMIUM.—Osmium is closely associated with iridium, from which it is separated by mercury. At first it is a black powder without metallic lustre; when further treated it becomes a white metal, not so brilliant as platinum, and is easily reduced to powder. When first obtained from the amalgam it is very combustible, and the mass burns away, being converted into the volatile oxide, or *osmic acid*. Five oxides of this metal are given, but osmic acid is the only one that is formed directly. Its density is 10.

RHODIUM.—Rhodium was obtained by Wollaston in the ore of platinum. An ore from Brazil contained 0·4 per cent., and an ore from another locality has yielded as much as 3 per cent. of rhodium. Rhodium, when fully prepared, is a white metal with a density of about 11. In a pure state it is not affected by acid, but when alloyed with platinum, bismuth, lead, or copper it dissolves with those metals. The solutions of the metal have a beautiful red colour, whence its name from *ῥόδον*, a rose.

RUTHENIUM.—A rare metal allied to the last.

The substances enumerated and described in the foregoing pages are distributed throughout the strata of the earth's crust, as given in the following table.

TABLE OF STRATA.

CAINOZOIC OR TERTIARY	*Recent.*		
	POST PLIOCENE.		
	NEWER PLIOCENE.		
	OLDER PLIOCENE.		
	MIOCENE.		
	EOCENE.		
			Lower, Middle, and Upper.
MESOZOIC OR SECONDARY	CRETACEOUS		Chalk. Upper Greensand. Gault. Lower Greensand.
	WEALDEN		Wealden. Purbeck Beds.
	OOLITE	Upper	Portland Oolite. Kimmeridge Clay.
		Middle	Coral Rag. Oxford Clay.
		Lower	Cornbrash. Forest Marble. Bath or Great Oolite. Stonesfield Slate. Inferior Oolite.
	LIAS		Upper Lias. Marlstone. Lower Lias.
	TRIAS		Rhaetic Beds. Keuper (New Red Marl). Bunter (New Red Sandstone).

TABLE OF STRATA—*continued.*

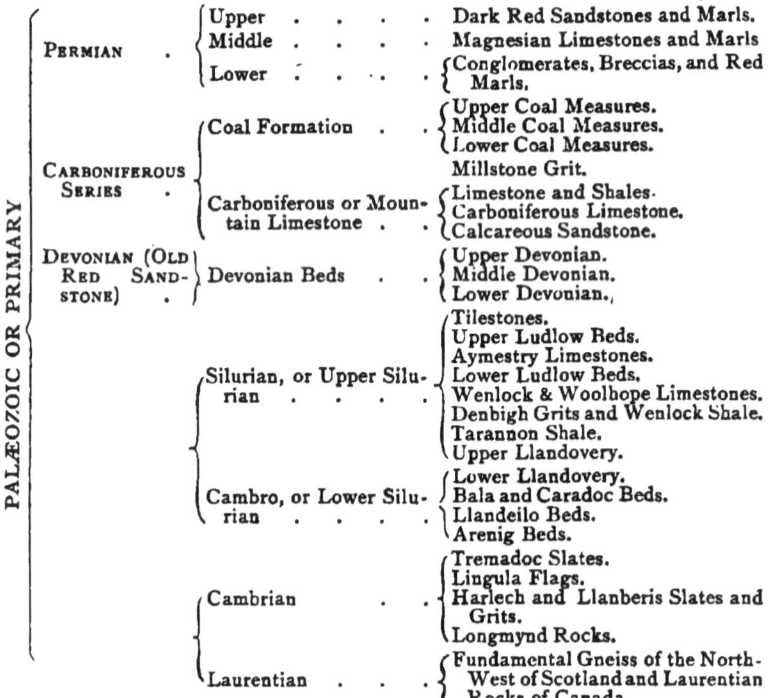

Era	System	Division	Strata
PALÆOZOIC OR PRIMARY	PERMIAN	Upper	Dark Red Sandstones and Marls.
		Middle	Magnesian Limestones and Marls
		Lower	Conglomerates, Breccias, and Red Marls.
	CARBONIFEROUS SERIES	Coal Formation	Upper Coal Measures. Middle Coal Measures. Lower Coal Measures.
			Millstone Grit.
		Carboniferous or Mountain Limestone	Limestone and Shales. Carboniferous Limestone. Calcareous Sandstone.
	DEVONIAN (OLD RED SANDSTONE)	Devonian Beds	Upper Devonian. Middle Devonian. Lower Devonian.
		Silurian, or Upper Silurian	Tilestones. Upper Ludlow Beds. Aymestry Limestones. Lower Ludlow Beds. Wenlock & Woolhope Limestones. Denbigh Grits and Wenlock Shale. Tarannon Shale. Upper Llandovery.
		Cambro, or Lower Silurian	Lower Llandovery. Bala and Caradoc Beds. Llandeilo Beds. Arenig Beds.
		Cambrian	Tremadoc Slates. Lingula Flags. Harlech and Llanberis Slates and Grits. Longmynd Rocks.
		Laurentian	Fundamental Gneiss of the North-West of Scotland and Laurentian Rocks of Canada.

It will be observed that the first six classes of minerals have a much wider range through time then have the distinctively metalliferous minerals. It may also be noticed generally that while the latter are more strictly confined to lodes, veins, and contact deposits, the former occur in beds, large deposits in beds, and in irregular accumulations in strata and in previously formed cavities.

INDEX.

PRINTED BY J. S. VIRTUE AND CO., LIMITED, CITY ROAD, LONDON.

www.ingramcontent.com/pod-product-compliance
Lightning Source LLC
Chambersburg PA
CBHW021404210326
41599CB00011B/1002

* 9 7 8 3 7 4 4 7 9 0 4 6 8 *